Biometrics

Identity Verification in a Networked World

A Wiley Tech Brief

Samir Nanavati
Michael Thieme
Raj Nanavati

Wiley Computer Publishing

John Wiley & Sons, Inc.

NEW YORK • CHICHESTER • WEINHEIM • BRISBANE • SINGAPORE • TORONTO

Publisher: Robert Ipsen
Editor: Margaret Eldridge
Editor: Adaobi Obi
Managing Editor: Micheline Frederick
Text Design & Composition: John Wiley Composition Services

Designations used by companies to distinguish their products are often claimed as trademarks. In all instances where John Wiley & Sons, Inc., is aware of a claim, the product names appear in initial capital or ALL CAPITAL LETTERS. Readers, however, should contact the appropriate companies for more complete information regarding trademarks and registration.

This book is printed on acid-free paper. ∞

Copyright © 2002 by Samir Nanavati. All rights reserved.

Published by John Wiley & Sons, Inc.

Published simultaneously in Canada.

No part of this publication may be reproduced, stored in a retrieval system or transmitted in any form or by any means, electronic, mechanical, photocopying, recording, scanning or otherwise, except as permitted under Sections 107 or 108 of the 1976 United States Copyright Act, without either the prior written permission of the Publisher, or authorization through payment of the appropriate per-copy fee to the Copyright Clearance Center, 222 Rosewood Drive, Danvers, MA 01923, (978) 750-8400, fax (978) 750-4744. Requests to the Publisher for permission should be addressed to the Permissions Department, John Wiley & Sons, Inc., 605 Third Avenue, New York, NY 10158-0012, (212) 850-6011, fax (212) 850-6008, E-mail: PERMREQ @ WILEY.COM.

This publication is designed to provide accurate and authoritative information in regard to the subject matter covered. It is sold with the understanding that the publisher is not engaged in professional services. If professional advice or other expert assistance is required, the services of a competent professional person should be sought.

ISBN: 0471-09945-7

Printed in the United States of America.

10 9 8 7 6 5 4 3 2 1

Wiley Tech Brief Series

Other titles in this series:

Optical Networking, by Deborah Cameron, ISBN 0471443689

Service Providers, by Joseph R. Matthews and Mary Helen Gillespie, ISBN 0471418188

Internet Appliances, by Ray Rischpater, ISBN 0471441112

PKI, by Tom Austin, ISBN 0471353809

The Wireless Application Protocol (WAP), by Steve Mann and Scott Sbihli, ISBN: 047139992

Palm Enterprise Applications, by Ray Rischpater, ISBN: 0471393797

Wireless Internet Enterprise Applications, by Chetan Sharma, ISBN: 0471383827

Application Service Providers, by Traver Gruen-Kennedy, ISBN: 0471394912

Cryptography and E-Commerce, by Jon Graff, ISBN: 0471405744

Enterprise Application Integration, by William Ruh, Francis X. Maginnis, William Brown, ISBN: 0471376418

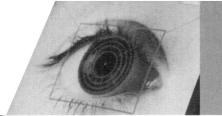

Contents

Introduction		xv
Acknowledgments		xix
Part One	**Biometric Fundamentals**	**1**
Chapter 1	**Why Biometrics?**	**3**
	Benefits of Biometrics versus Traditional Authentication Methods	3
	Increased Security	4
	Increased Convenience	5
	Increased Accountability	5
	Benefits of Biometrics in Identification Systems	6
	Fraud Detection	6
	Fraud Deterrence	6
	Conclusion: Evaluating the Benefits of Biometric Systems	6
Chapter 2	**Key Biometric Terms and Processes**	**9**
	Definitions	9
	Discussion: Verification and Identification	12
	When are Verification and Identification Appropriate?	13
	Between Verification and Identification: *1:Few*	14
	Logical versus Physical Access	14

		How Biometric Matching Works	15
		Enrollment and Template Creation	16
		Templates	18
		Biometric Matching	20
		Conclusion	22
	Chapter 3	**Accuracy in Biometric Systems**	**23**
		False Match Rate	24
		Importance of FMR	25
		When Are False Matches Acceptable?	25
		Single False Match Rate versus System False Match Rate	26
		False Match Rate in Large-Scale Identification Systems	27
		False NonMatch Rate	27
		Importance of FNMR	28
		Changes in a User's Biometric Data	28
		Changes in User Presentation	29
		Changes in Environment	29
		Real-World False Nonmatch Rates	31
		Single FNMR versus System FNMR	31
		False Nonmatch Rates in Large-Scale Identification Systems	32
		Failure-to-Enroll (FTE) Rate	33
		Importance of FTE Rates	35
		Relation of Failure-to-Enroll to False Nonmatch Rate	35
		FTE across Different Populations	35
		Single FTE Rates versus System FTE Rates	37
		FTE in Large-Scale Identification Systems	38
		Derived Metrics	38
		Equal Error Rate (EER)	38
		Ability-to-Verify (ATV) Rate	39
		Conclusion: Biometric Technologies from an Accuracy Perspective	40
Part 2		**Leading Biometric Technologies: What You Need to Know**	**43**
	Chapter 4	**Finger-Scan**	**45**
		Components	46
		How Finger-Scan Technology Works	48
		Image Acquisition	48
		Image Processing	50
		Location of Distinctive Characteristics	51
		Template Creation	52
		Template Matching	52

	Competing Finger-Scan Technologies	54
	Optical Technology	54
	Silicon Technology	54
	Ultrasound Technology	55
	Feature Extraction Methods: Minutiae versus Pattern Matching	55
	Finger-Scan Deployments	56
	Finger-Scan Strengths	58
	Proven Technology Capable of High Levels of Accuracy	58
	Range of Deployment Environments	58
	Ergonomic, Easy-to-Use Devices	59
	Ability to Enroll Multiple Fingers	59
	Finger-Scan Weaknesses	59
	Inability to Enroll Some Users	59
	Performance Deterioration over Time	60
	Association with Forensic Applications	60
	Need to Deploy Specialized Devices	60
	Finger-Scan: Conclusion	61
Chapter 5	**Facial-Scan**	**63**
	Components	64
	How Facial-Scan Technology Works	65
	Image Acquisition	65
	Image Processing	66
	Distinctive Characteristics	67
	Template Creation	68
	Template Matching	68
	Competing Facial-Scan Technologies	69
	Eigenface	69
	Feature Analysis	70
	Neural Network	71
	Automatic Face Processing	71
	Facial-Scan Deployments	72
	Facial-Scan Strengths	72
	Ability to Leverage Existing Equipment and Imaging Processes	73
	Ability to Operate without Physical Contact or User Complicity	73
	Ability to Enroll Static Images	73
	Facial-Scan Weaknesses	74
	Acquisition Environment Effect on Matching Accuracy	74
	Changes in Physiological Characteristics That Reduce Matching Accuracy	74
	Potential for Privacy Abuse Due to Noncooperative Enrollment and Identification	75
	Facial-Scan: Conclusion	75

Chapter 6 Iris-Scan   77
Components   78
How It Works   79
- Image Acquisition   79
- Image Processing   80
- Distinctive Features   80
- Template Generation   81
- Template Matching   82

Deployments   82
Iris-Scan Strengths   83
- Resistance to False Matching   83
- Stability of Characteristic over Lifetime   84
- Suitability for Logical and Physical Access   84

Iris-Scan Weaknesses   84
- Difficulty of Usage   85
- False Nonmatching and Failure-to-Enroll   85
- User Discomfort with Eye-Based Technology   85
- Need for a Proprietary Acquisition Device   86

Iris-Scan: Conclusion   86

Chapter 7 Voice-Scan   87
Components   88
How It Works   88
- Data Acquisition   89
- Data Processing   90
- Distinctive Features   90
- Template Creation   91
- Template Matching   92

Deployments   92
Voice-Scan Strengths   93
- Ability to Leverage Existing Telephony Infrastructure   94
- Synergy with Speech Recognition and Verbal Account Authentication   94
- Resistance to Imposters   94
- Lack of Negative Perceptions Associated with Other Biometrics   95

Voice-Scan Weaknesses   95
- Effect of Acquisition Devices and Ambient Noise on Accuracy   95
- Perception of Low Accuracy   96
- Lack of Suitability for Today's PC Usage   96
- Large Template Size   96

Voice-Scan: Conclusion   97

Chapter 8	**Other Physiological Biometrics**		**99**
	Hand-Scan		99
		Components	100
		How It Works	101
		Deployments	102
		Hand-Scan Strengths	103
		Hand-Scan Weaknesses	105
		Conclusion	106
	Retina-Scan		106
		Components	107
		How It Works	107
		Deployments	110
		Retina-Scan Strengths	110
		Retina-Scan Weaknesses	111
		Conclusion	112
	Automated Fingerprint Identification Systems (AFIS)		114
		Components	114
		How It Works	116
		Deployments	119
		How AFIS and Finger-Scan Differ	120
		Conclusion	121
Chapter 9	**Other Leading Behavioral Biometrics**		**123**
	Signature-Scan		123
		Components	124
		How It Works	125
		Deployments	127
		Signature-Scan Strengths	128
		Signature-Scan Weaknesses	130
		Conclusion	131
	Keystroke-Scan		132
		Components	134
		How It Works	134
		Keystroke-Scan Strengths	136
		Keystroke-Scan Weaknesses	137
		Conclusion	139
Part 3	**Biometric Applications and Markets**		**141**
Chapter 10	**Categorizing Biometric Applications**		**143**
	Defining the Seven Biometric Applications		144
	Capacities in Which Individuals Use Biometric Systems		147

		Introduction to IBG's Biometric Solution Matrix	147
		How Urgent Is the Authentication Problem That Biometrics Are Solving?	148
		What Is the Scope of the Authentication Problem That Biometrics Are Solving?	149
		How Well Can Biometrics Solve the Authentication Problem?	149
		Are Biometrics the Only Possible Authentication Solution?	149
		How Receptive Are Users to Biometrics as an Authentication Solution?	149
Chapter 11	**Citizen-Facing Applications**		**151**
	Criminal Identification		152
		Today's Criminal Identification Applications	152
		Future Criminal Identification Trends	152
		Related Biometric Technologies and Vertical Markets	154
		Cost to Deploy Biometrics in Criminal Identification	156
		Conclusion	156
	Citizen Identification		157
		Typical Applications	157
		Future Trends in Citizen Identification	158
		Related Biometric Technologies and Vertical Markets	161
		Cost to Deploy Biometrics in Citizen Identification	161
		Issues Involved in Deployment	162
		Conclusion	164
	Surveillance		164
		Today's Surveillance Applications	164
		Future Trends in Surveillance	165
		Related Biometric Technologies and Vertical Markets	167
		Cost to Deploy Biometrics in Surveillance	167
		Issues Involved in Deployment	168
		Conclusion	169
Chapter 12	**Employee-Facing Applications**		**171**
	PC/Network Access		171
		Today's PC/Network Access Applications	172
		Future Trends in PC/Network Access	173
		Related Biometric Technologies and Vertical Markets	177
		Costs to Deploy Biometrics in PC/Network Access	177
		Issues Involved in Deployment	178
		Conclusion	180

| | | Contents | xi |

	Physical Access/Time and Attendance	180
	Today's Physical Access/Time and Attendance Applications	181
	Future Physical Access/Time and Attendance Trends	182
	Related Biometric Technologies and Vertical Markets	183
	Costs to Deploy Biometrics in Physical Access/Time and Attendance	183
	Issues Involved in Deployment	186
	Conclusion	187

Chapter 13 Customer-Facing Applications 189

E-Commerce/Telephony	190
Today's E-Commerce/Telephony Applications	190
Future E-Commerce/Telephony Trends	191
Related Biometric Technologies and Vertical Markets	193
Costs to Utilize Biometrics in E-Commerce/Telephony	193
Issues Involved in Deployment	198
Conclusion	201
Retail/ATM/Point of Sale	201
Today's Retail/ATM/Point-of-Sale Applications	201
Future Retail/ATM/Point-of-Sale Trends	203
Related Biometric Technologies and Vertical Markets	206
Cost to Deploy Biometrics in Retail/ATM/POS	206
Issues Involved in Deployment	207
Conclusion	208

Chapter 14 Biometric Vertical Markets 209

Five Primary Biometric Vertical Markets	210
Law Enforcement	211
Technologies Used in Law Enforcement	211
Typical Law Enforcement Deployments	211
Conclusion	213
Government Sector	214
Technologies Used in the Government Sector	214
Typical Government-Sector Deployments	215
Conclusion	220
Financial Sector	220
Technologies Used in the Financial Sector	221
Typical Financial-Sector Deployments	222
Conclusion	225

		Healthcare	225
		Technologies Used in Healthcare	226
		Typical Healthcare Deployments	226
		Conclusion	228
		Travel and Immigration	228
		Technologies Used in Travel and Immigration	229
		Typical Travel and Immigration Deployments	229
		Conclusion	231
		Additional Biometric Verticals	231
		Conclusion	233
Part 4		**Privacy and Standards In Biometric System Design**	**235**
	Chapter 15	**Assessing the Privacy Risks of Biometrics**	**237**
		Biometric Deployments on a Privacy Continuum	238
		Privacy Concerns Associated with Biometric Deployments	239
		Informational Privacy	239
		Personal Privacy	243
		Privacy-Sympathetic Qualities of Biometric Technology	244
		Defining Application-Specific Privacy Risks: The BioPrivacy Impact Framework	246
		Overt versus Covert	246
		Opt-in versus Mandatory	248
		Verification versus Identification	249
		Fixed Duration versus Indefinite Duration	250
		Public Sector versus Private Sector	250
		Citizen, Employee, Traveler, Student, Customer, Individual	251
		User Ownership versus Institutional Ownership of Biometric Data	253
		Personal Storage versus Storage in Template Database	253
		Behavioral versus Physiological Biometric Technology	254
		Template Storage versus Identifiable Data Storage	255
		Conclusion	255
		BioPrivacy Technology Risk Ratings	256
	Chapter 16	**Designing Privacy-Sympathetic Biometric Systems**	**259**
		BioPrivacy Best Practices: Scope and Capabilities	260
		Limit System Scope	260
		Do Not Use Biometrics as a Unique Identifier	260

	Limit Retention of Biometric Information	261
	Evaluate a System's Potential Capabilities	263
	Limit Storage of Identifiable Biometric Data	264
	Limit Collection and Storage of Extraneous Information	264
	Make Provisions for System Termination	265
IBG BioPrivacy Best Practices: Data Protection		265
	Use Security Tools and Access Policies to Protect Biometric Information	265
	Protect Postmatch Decisions	266
	Limit System Access	266
	Implement Logical and Physical Separations between Biometric and Nonbiometric Data	267
BioPrivacy Best Practices: User Control of Personal Data		267
	Make System Usage Voluntary and Allow for Unenrollment	267
	Enable Anonymous Enrollment and Verification	268
	Provide Means of Correcting and Accessing Biometric-Related Information	268
IBG BioPrivacy Best Practices: Disclosure, Auditing, and Accountability		269
	Make Provisions for Third-Party Auditing and Oversight	269
	Hold Operators Accountable for System Use and Misuse	269
	Fully Disclose Audit Findings	270
	Disclose the System Purpose and Objectives	270
	Disclose When Individuals May Be Enrolled in a Biometric System	271
	Disclose When Individuals May Be Verified in a Biometric System	271
	Disclose Whether Enrollment Is Optional or Mandatory	271
	Disclose Enrollment, Verification, and Identification Processes	272
	Disclose Policies and Protections in Place to Ensure Privacy of Biometric Information	272
Biometrics at the Super Bowl: An IBG BioPrivacy Assessment		273
Conclusion		276
Chapter 17	**Biometric Standards**	**277**
Why Standards?		277
Application Programming Interfaces		278
	BioAPI	279
	BAPI	280
	Deployers, Developers, and Biometric APIs	280

File Format	281
Information Security for Financial Services	281
Additional Efforts	282
Fingerprint Template Interoperability	282
CDSA/HRS	284
Conclusion	284
Index	**285**

Introduction

Authentication is a fundamental component of human interaction with computers. Traditional means of authentication, primarily passwords and personal identification numbers (PINs), have until recently dominated computing, and are likely to remain essential for years to come. However, stronger authentication technologies, capable of providing higher degrees of certainty that a user is who he or she claims to be, are becoming commonplace. Biometrics are one such strong authentication technology.

Biometric technologies as we know them today have been made possible by explosive advances in computing power and have been made necessary by the near universal connectedness of computers around the world. The increased perception of data and information as near equivalents of currency, in conjunction with the opportunities for access provided by the Internet, is a paradigm shift with significant repercussions for authentication. If data is currency, then server-based or local hard drives are our new vaults, and information-rich companies will be held responsible for their security. Because of this, passwords and PINs are nearing the end of their life cycle for many applications.

Since early 1999, four factors (reduced cost, reduced size, increased accuracy, and increased ease of use) have combined to make biometrics an increasingly feasible solution for securing access to computers and networks. But biometrics are much more than a replacement for passwords. Millions of people around the world use biometric technology in applications as varied as time and attendance, voter registration, international travel, and benefits dispersal. Depending on the application, biometrics can be used for security, for convenience, for fraud reduction, even as an empowering technology.

This book teaches you the fundamentals of leading biometric technologies: how they work, their strengths and weaknesses, where they can be effectively deployed. It helps you understand how biometrics are associated with technologies such as public key infrastructure (PKI) and smart cards. It defines how biometric deployments can be privacy enhancing or privacy invasive. It dispels various myths surrounding biometric technology. Finally, it provides guidelines for successful deployment of biometrics in today's enterprise environment.

How This Book Is Organized

There are many misconceptions about biometric technology that have taken hold in the general public. In order to provide you with a solid understanding of the technology, the industry, the applications, and the challenges that define biometrics, this book is divided into four parts.

Part One: Biometric Fundamentals

Part One provides you with a detailed understanding of the central concepts involved in biometrics, including reasons why the technology is deployed, how the technology operates, the key processes involved in biometric authentication, and how accuracy is defined in biometric systems.

Chapter 1, "Why Biometrics?," discusses why biometrics are of such interest to institutions looking to authenticate employees, customers, and citizens. This chapter also details the benefits that biometric technologies can provide when deployed correctly. Chapter 2, "Key Biometric Terms and Processes," digs deeper into biometric functionality, defining and explaining the variety of terms, both technical and nontechnical, used in the biometric industry. This chapter also discusses the biometric concepts essential to understanding how and why biometrics are deployed today. Chapter 3, "Accuracy in Biometric Systems," defines the categories used to determine how well biometric systems work and examines why most discussions of biometric accuracy are highly misleading.

Part Two: Leading Biometric Technologies

Part Two provides detailed discussions of leading biometric technologies. The information in Part Two is based on real-world experience in deploying and testing systems in operational environments. For each biometric discipline, Part Two addresses in full topics such as system components, data acquisition, template generation and matching, deployments, and strengths and weaknesses. Technologies discussed include finger-scan, facial-scan, iris-scan, retina-scan, signature-scan, keystroke-scan, hand-scan, AFIS, and biometric middleware.

Part Three: Biometric Applications and Markets

Part Three discusses the applications in which biometrics are used and the markets and industries in which they are most effectively deployed. Biometric solutions have emerged in a range of applications, including Citizen ID, Network/PC Access, Surveillance, and e-Commerce/Telephony. These applications can differ substantially in terms of technology selection, privacy impact, and performance requirements. Furthermore, biometrics have emerged as viable solutions in a handful of vertical markets, including government and financial sectors.

Part Four: Privacy and Standards In Biometric System Design

Part Four discusses critical factors related to standards and privacy that institutions must consider when designing, deploying, and maintaining biometric systems for customers, employees, or citizens. Privacy has long been a major issue in biometrics; the emergence of a framework for evaluating the privacy impact of biometric technologies and deployments should help address this central problem. In addition, the emergence of biometric standards has reduced the levels of risk involved in deploying biometric systems.

Who Should Read This Book

This book provides public- and private-sector professionals with a real-world understanding of biometric technologies and applications, helping them make informed decisions on the role that biometrics can play in their organization. No background in biometrics is required, but those with biometrics experience will benefit from the detailed discussion of concepts not often presented from a deployer's perspective, such as accuracy, privacy, and technology strengths and weaknesses. Although some of the discussions may go into detail about biometric processes or functions, the book is not intended to be highly technical.

Those new to biometrics are strongly encouraged to read from the beginning, as an understanding of biometrics requires familiarity with the reasons why biometrics are deployed and the key terms involved in biometrics. Those more experienced with biometrics can begin with the technologies of interest to them and move into privacy and system design. However, even those with biometrics experience are encouraged to read the entire book(many basic concepts in the biometrics industry have often been poorly defined or overlooked, a problem that this book addresses.

Looking Forward

Biometric technology has emerged as a viable solution for a range of applications where a person's identity must be verified or determined. No longer a science-fiction solution, biometrics are being deployed to solve security problems, to help companies generate revenues, and to protect personal information. The challenge is to ensure that biometrics are used intelligently and responsibly as the technology moves into the mainstream; the repercussions of unintelligent and irresponsible use could be severe.

Acknowledgments

The authors would like to express their appreciation to those whose diligent efforts have been instrumental to the growth of the biometric industry, and to those who have helped foster a greater understanding of the potential impact of biometrics on today's world:

Cathy Tilton, William Saito, Geoff Slagle, Barry Steinhardt, Peter Hope-Tindall, Colleen Madigan, Joseph Atick, Bill Voltmer, Jeff Stapleton, Paul Reid, Scott Moody, John Ticer, Karl Ware, Vance Bjorn, Erik Bowman, Mitch Tarr, Fernando Podio, Jeff Dunn, Dr. Ann Cavoukian, John Woodward, Dr. Jim Wayman, Dr. Doug McGovern, Peter Higgins, Bill Rogers, Bill Spence, John Harris, Rick Pratt, Dave Troy, Dr. Anil Jain, Oz Pieper, Ron Beyner, Walter Hamilton, Tom Hopper, Mike Garris, Ed German, Colin Soutar, Naeem Zafar, Dr. Bridgette Wirtz, Dave Mintie, Astrid Albrecht, Tony Mansfield, Dr. John Daugman, Denny Carlton, Tim Ruggles, Norm Hughes, Sid Lieberman, Gary Roethenbaugh, Steve Borza, Dr. Larry O'Gorman, Tom Colatosti, Dr. John Schneider, Dennis Quiggle, Jeff Poulson, Allen Ganz, David Hertz, Dae Won Im, Christer Bergman, Fabio Righi, Dr. Jonathan Phillips, Duane Blackburn, P.J. Bulger, and Jean-Marc Suchier.

Special thanks go to: J. Shoor, L. Corbett, J. Goodman, C. Russo, A. Carr, L. Wall, A. Garay, C. Connors, G. Gruber, B. Chen, J. Granato, R. Rasmussen, J. Goding, B. Hyslop, B. Aucoin, J. Homme, C. Thompson, R. Pollard, D. Fuchs, R. Hoyte, E. Olsen, J. Gee, A. Backenroth, M. Hanselman, T. Dorren, J. Medicus, K. Nessman, P. Anther, S. Walsh, M. Nazar, D. Ziemke, C. Nagy, M. Goldberg, I. Beyah, and M. Spivey.

The authors would like to recognize the research, consulting, integration and support professionals of International Biometric Group for their hard work and dedication. Without their creativity, integrity, and commitment,

International Biometric Group could not have grown into its role as the internationally recognized leader in the biometric industry. Specifically, thanks go to the following IBG personnel: A. Green, B. Nelson, D. Stipisic, D. Most, E. Chorbajain, K. Hunt, G. Glasser, J. Hong, M. Mak, and M. Duitz.

Most of all, the authors would like to personally thank Yasmin, Harit, Karen, Margaret, Charles, and Heather for their support and encouragement.

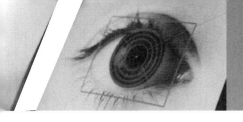

PART ONE

Biometric Fundamentals

Deploying biometrics effectively requires a solid understanding of the fundamentals of the technology, including why it is deployed, how it works, and how performance and accuracy are measured. While leading biometric technologies vary in complexity, capabilities, and performance, there are a number of elements shared by all biometric technologies. Template generation, matching and error rates, and enrollment processes are among the many concepts central to biometric that have a significant impact on system design, deployment costs, and individual privacy.

The chapters in this section provide potential biometric deployers with a comprehensive understanding of the fundamentals of biometrics, including the reasons why biometrics are deployed, the key terms used in the industry, and the essentials of biometrics technology accuracy and performance.

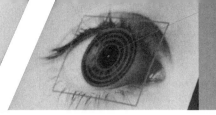

CHAPTER 1

Why Biometrics?

Biometric technology is used for dozens of types of applications, ranging from modest (providing time and attendance functionality for a small company) to expansive (ensuring the integrity of a 10 million-person voter registration database). Depending on the application, the benefit of using or deploying biometrics may be increased security, increased convenience, reduced fraud, or delivery of enhanced services. In some applications, the biometric serves only as a deterrent; in others, it is central to system operation. Regardless of the rationale for deploying biometrics, there are two common elements:

1. The benefits of biometric usage and deployment are derived from having a high degree of certainty regarding an individual's identity.
2. The benefits lead directly or indirectly to cost savings or to reduced risk of financial losses for an individual or institution.

Benefits of Biometrics versus Traditional Authentication Methods

The most frequently used authentication technologies are passwords and PINs. They secure access to personal computers (PCs), networks, and applications; control entry to secure areas of a building; and authorize automatic teller machine (ATM) and debit transactions. Handheld tokens (such as cards and key fobs) have replaced passwords in some higher-security applications.

However, passwords, PINs, and tokens have a number of problems that call into question their suitability for modern applications, particularly high-security applications such as access to online financial accounts or medical data. What benefits do biometrics provide compared to these authentication methods?

Increased Security

Biometrics can provide a greater degree of security than traditional authentication methods, meaning that resources are accessible only to authorized users and are kept protected from unauthorized users. Passwords and PINs are easily guessed or compromised; tokens can be stolen. Many users select obvious words or numbers for password or PIN authentication, such that an unauthorized user may be able to break into an account with little effort. In addition, many users write passwords in conspicuous places, especially as the number of passwords users must manage continually increases. *Good* passwords—long passwords with numbers and symbols—are too difficult to remember for most users and are rarely enforced.

By contrast, biometrics data cannot be *guessed* or *stolen* in the same fashion as a password or token. Although some biometric systems can be broken under certain conditions, today's biometric systems are highly unlikely to be fooled by a picture of a face, an impression of a fingerprint, or a recording of a voice. This assumes, of course, that the imposter has been able to gather these physiological characteristics—unlikely in most cases.

In systems where the biometric authentication releases passwords (leveraging the existing username-password infrastructure), the user or administrator can create longer and more complex passwords than would be feasible without biometrics.

Passwords, PINs, and tokens can also be shared, which increases the likelihood of malicious or unaccountable use. In many enterprises, a common password is shared among administrators to facilitate system administration. Unfortunately, because there is no certainty as to who is using a shared password or token—or whether the user is even authorized—security and accountability are greatly reduced. Being based on distinctive characteristics, biometric data cannot be shared in this fashion, although in some systems two users can choose to share a joint bank account by each enrolling a fingerprint.

Although there are a number of security issues involved in biometric system usage that must be addressed through intelligent system design, the level of security provided by most biometric systems far exceeds the security provided by passwords, PINs, and tokens. In later chapters, as we discuss the business case for biometrics in specific applications, we will find that security frequently plays a large role in an enterprise's decision to deploy biometrics.

Increased Convenience

One of the reasons passwords are kept simple (and are then subject to compromise) is that they are easily forgotten. As computer users are forced to manage more and more passwords, the likelihood of passwords being forgotten increases, unless users choose a universal password, reducing security further. Tokens and cards can be forgotten as well, though keeping them attached to keychains reduces this risk.

Because biometrics are difficult if not impossible to forget, they can offer much greater convenience than systems based on remembering multiple passwords or on keeping possession of an authentication token. For PC applications in which a user must access multiple resources, biometrics can greatly simplify the authentication process—the biometric replaces multiple passwords, in theory reducing the burden on both the user and the system administrator. Applications such as point-of-sale transactions have also begun to see the use of biometrics to authorize purchases from prefunded accounts, eliminating the need for cards.

Biometric authentication also allows for association of higher levels of rights and privileges with a successful authentication. Highly sensitive information can more readily be made available on a biometrically protected network than on one protected by passwords. This can increase user and enterprise convenience, as users can access otherwise protected information without the need for human intervention.

Increased Accountability

Given the increased awareness of security issues in the enterprise and in customer-facing applications, the need for strong auditing and reporting capabilities has grown more pronounced. Using biometrics to secure computers and facilities eliminates phenomena such as buddy-punching and provides a high degree of certainty as to what user accessed what computer at what time. Even if the auditing and reporting capabilities of a system are rarely used, the fact that they exist often serves as an effective deterrent.

The benefits of security, convenience, and accountability apply primarily to enterprises, corporations, and home users; in addition, they describe the rationale for biometric *verification*. The benefits of biometric *identification*, especially on a large scale, differ substantially.

Benefits of Biometrics in Identification Systems

In identification systems, biometrics can still be used for security, convenience, and accountability, especially when they are deployed to a modest number of users. However, identification systems are more often deployed in large-scale environments, anywhere from tens of thousands to tens of millions of users. In these applications, biometric identification is not replacing passwords or PINs—it is providing new types of fraud-reducing functionality.

Fraud Detection

Identification systems are deployed to determine whether a person's biometric information exists more than once in a database. By locating and identifying individuals who have already registered for a program or service, biometrics can reduce fraud. In a public benefits program, for example, a person may be able to register under multiple identities using fraudulent documentation. A person can also obtain fraudulent identification such as a driver's license. Without biometrics, there is no way to be certain that a person is not registered under a different identity.

Fraud Deterrence

Perhaps even more than fraud detection, fraud deterrence is a primary benefit in large-scale identification systems. It can be difficult to return a highly certain match against millions of existing biometric records: In some cases, the error rates in large-scale identification systems can run into the single digits (much higher than would be acceptable in verification applications). However, the very presence of the biometric provides a benefit, as it dissuades people who might otherwise be prone to attempt multiple registration. If the presence of biometric identification technology can deter individuals from attempting to enroll multiple times in a public benefit or driver's license system, then the public agency has saved money and ensured the integrity of its records. In the absence of biometrics, there is no efficient way of identifying duplicate applicants or registrants, and it is therefore difficult to deter such applications.

Conclusion: Evaluating the Benefits of Biometric Systems

Notwithstanding the benefits of biometric technology, biometrics are not suitable for every application and user, and in some cases biometric authentication is simply the wrong solution. One of the major challenges facing the biometric industry is defining those environments in which biometrics provide the strongest benefit to individuals and institutions, and then demonstrating that the benefits of deployment outweigh the risks and costs. Over time, the increased effectiveness and affordability of biometric technologies has continually broadened the range of applications in which biometrics operate effectively.

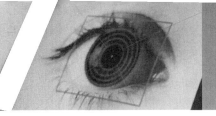

CHAPTER 2

Key Biometric Terms and Processes

Because *biometrics* refers to such a broad range of technologies, systems, and applications, it is essential to discuss the terminology, classifications, and unique processes that define biometrics.

Definitions

In the following section we present definitions of key biometric terms and discuss key facets of these definitions.

Biometrics is the automated use of physiological or behavioral characteristics to determine or verify identity.

Several aspects of this definition require elaboration.

Automated use. Behavioral and physiological characteristics are regularly used to *manually* verify or determine identity—this is something that humans do every day when we greet friends or check an ID card. Biometric technologies, by contrast, are automated—computers or machines are used to verify or determine identity through behavioral or physiological characteristics. Because the process is automated, biometric authentication generally requires only a few seconds, and biometric systems are able to compare thousands of records per second. A forensic investigator performing a visual match against an ink fingerprint is not performing biometric authentication. By contrast, a system wherein a user places his or her finger on a reader and a match/no-match decision is rendered in real time is performing biometric authentication.

Physiological or behavioral characteristics. Biometrics are based on the measurement of distinctive physiological and behavioral characteristics. Finger-scan, facial-scan, iris-scan, hand-scan, and retina-scan are considered physiological biometrics, based on direct measurements of a part of the human body. Voice-scan and signature-scan are considered behavioral biometrics; they are based on measurements and data derived from an action and therefore *indirectly* measure characteristics of the human body. The element of time is essential to behavioral biometrics—the characteristic being measured is tied to an action, such as a spoken or signed series of words, with a beginning and an end. The physiological/behavioral classification is a useful way to view the types of biometric technologies, because certain performance- and privacy-related factors often differ between the two types of biometrics. However, the behavioral/physiological distinction is slightly artificial. Behavioral biometrics are based in part on physiology, such as the shape of the vocal cords in voice-scan or the dexterity of hands and fingers in signature-scan. Physiological biometric technologies are similarly informed by user behavior, such as the manner in which a user presents a finger or looks at a camera.

Determine versus verify. Determining versus verifying identity represents a fundamental distinction in biometric usage. Some biometric systems can determine the identity of a person from a biometric database without that person first claiming an identity. The traditional use of fingerprints in crime investigations—searching stored records of fingerprints in order to find a match—is an identification deployment. Identification systems stand in contrast to verification systems, in which a person claims a specific identity and the biometric system either confirms or denies that claim. Accessing a network is normally a verification event—the user enters an ID and verifies that he or she is the proper user of that ID by entering a password or biometric. Identification and verification systems differ substantially in terms of privacy, performance, and integration into existing systems. This critical distinction is discussed in more detail in following sections (see also Figure 2.1).

Identity. Identity is often misunderstood in the context of biometrics, where a distinction must be drawn between an individual and an identity. An individual is a singular, unique entity—colloquially, a person—but an individual can have more than one identity. For example, John Doe might have an

Key Biometric Terms and Processes 11

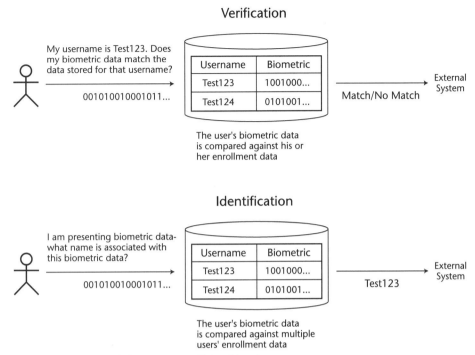

Figure 2.1 Determining versus verifying identity.

email identity of jdoe@sample.com and a *work* identity of John L. Doe. A person who has registered 10 fingers in a biometric system could be viewed as 10 separate identities, not as one individual, by the biometric system. This identity distinction is important because it establishes limits on the type of certainty that a biometric system can provide. It can also have significant bearing on biometrics and privacy. Biometric identity verification and determination are only as strong as the initial association of a biometric with an individual. A user who enrolls in a biometric system under a false identity will continue to have this false identity verified with every successful biometric match.

> **NOTE**
> *Biometric* can be used as a noun when referring to a single technology: "Finger-scan is a commonly used biometric." *Biometric* can also be used as an adjective: "A biometric system uses integrated hardware and software to conduct identification or verification."

Discussion: Verification and Identification

Perhaps the most fundamental distinction in biometrics is between verification and identification. Nearly all aspects of biometrics—performance, benefits and risks of deployment, privacy impact, and cost—differ when moving between these two types of systems.

Verification systems answer the question, *"Am I who I claim to be?"* by requiring that a user claim an identity in order for a biometric comparison to be performed. After a user claims an identity, he or she provides biometric data, which is then compared against his or her enrolled biometric data. Depending on the type of biometric system, the identity that a user claims might be a Windows username, a given name, or an ID number; the answer returned by the system is *match* or *no match*. Verification systems can contain dozens, thousands, or millions of biometric records, but are always predicated on a user's biometric data being matched against only his or her own enrolled biometric data. Verification is often referred to as *1:1 (one-to-one)*. The process of providing a username and biometric data is referred to as *authentication*.

Identification systems answer the question, *"Who am I?"* and do not require that a user claim an identity before biometric comparisons take place. The user provides his or her biometric data, which is compared to data from a number of users in order to find a match. The answer returned by the system is an identity such as a name or ID number. Identification systems can contain dozens, thousands, or millions of biometric records. Identification is often referred to as *1:N (one-to-N* or *one-to-many)*, because a person's biometric information is compared against multiple (*N*) records.

Within identification systems there is a further distinction between positive and negative. Positive identification systems are designed to find a match for a user's biometric information in a database of biometric information. A typical positive identification system would be a prison release program where individuals do not enter an ID number or use a card, but provide biometric data and are located within an inmate database. The anticipated result of a search in a positive identification system is a match. Negative identification systems, by contrast, are designed to ensure that a person's biometric information is *not* present in a database. This prevents people from enrolling twice in a system and is often used in large-scale public benefits programs in which users attempt to enroll multiple times to gain benefits under different names. Although the underlying biometric matching technology may be very similar to that of positive identification, the anticipated result of a search in a negative identification system is a nonmatch.

Identification systems with more than approximately 100,000 users are considered large-scale identification systems. Large-scale identification systems generally differ substantially from smaller-scale identification systems, especially in accuracy and response time, to the point where they effectively become a qualitatively different type of biometric technology.

Only certain biometric technologies are capable of performing identification, including finger-scan, iris-scan, retina-scan, and to a lesser degree facial-scan. Signature-scan, voice-scan, and hand-scan are incapable of identification, because the physiological and behavioral characteristics upon which they are based are not sufficiently distinctive.

When Are Verification and Identification Appropriate?

It is rare that an organization facing a specific problem will find itself deciding whether to deploy an identification or a verification system. Instead, certain applications naturally lend themselves to verification, and others require identification. PC and network security generally employ verification systems; access to buildings and rooms can be effective with either identification or verification systems, though verification is predominant; and large-scale public benefits programs generally utilize identification systems. To deploy an effective biometric system, you must understand and be prepared to evaluate the strengths and weaknesses of each system type in relation to your business and security needs.

Verification systems are generally faster and more accurate than identification systems. Instead of performing hundreds of comparisons against enrolled users, they need only match a person's data against his or her existing data. This requires less computing power and decreases the likelihood that the system will match an unauthorized user. Nearly all verification systems can render a match/no-match decision within less than one second. Verification systems, of course, cannot determine whether a given person is present in a database more than once.

Identification systems require more computational power than verification systems, because more comparisons take place before a match occurs—in some cases, millions of matches. In addition, there are more opportunities for an identification system to err, because many more matches must be conducted.

As a rule, identification systems are deployed when verification simply does not make sense (to eliminate duplicate enrollments, for example). Although the idea of performing identification in desktop environments may be appealing, the associated reduction in speed and accuracy generally outweigh these modest benefits of eliminating a username or ID.

Between Verification and Identification: *1:Few*

A modest type of identification known as *1:few (one-to-few)* involves an identification search against a small number of users. Biometric applications such as local PC access and low-security physical access are occasionally deployed in 1:few mode. A user provides biometric data before claiming an identity, and a search is conducted against a small set of enrolled users. One-to-few systems may have search capabilities to only 5 to 10 users or may be capable of searching up to 100 users—there is no official line separating 1:many from 1:few applications.

Logical versus Physical Access

Once a biometric system has determined or verified an identity, what happens? The answer depends on the purpose for which the system is deployed. Biometric systems, and in many ways the entire biometric industry, can be segmented according to the purposes for which verification and identification are being performed. The two primary uses for a biometric system are physical access and logical access.

Physical access systems monitor, restrict, or grant movement of a person or object into or out of a specific area. Most physical access implementations involve entry into a room or building: bank vaults, server rooms, control towers, or any location to which access is restricted. Time and attendance are a common physical access application, combining access to a location with an audit of when the authentication occurred. Physical access can also entail accessing equipment or materials, such as opening a safe or starting an automobile, although most of these applications are still speculative. When used in physical access systems, biometrics replace or complement keys, access cards, PIN codes, and security guards.

Logical access systems monitor, restrict, or grant access to data or information. Logging into a PC, accessing data stored on a network, accessing an account, or authenticating a transaction are examples of logical access. Biometrics replace or complement passwords, PINs, and tokens in logical access systems.

The core biometric functionality—acquiring and comparing biometric data—is often identical in physical and logical access systems. The same finger-scan algorithm and reader, for example, can be used for both desktop and doorway applications. What changes between the two is the external system into which the biometric functionality is integrated. In both physical and logical access systems, the biometric functionality is integrated into a larger system (be it a door control system, for example, or an operating system). The biometric match effects a result such as at the opening of a door or access to an operating system.

Because of the tremendous value of information stored on corporate networks and the transaction value of business-to-business (B2B) and business-to-consumer (B2C) e-commerce, the biometric industry views logical access as a much more lucrative industry segment in the long run than physical access. The number of times an individual needs to provide authentication to a PC in a given day might be 20 or 30, while the instances of physical access authentication are less frequent and generally entail less value. The value of information and other intangible assets continually increases as more data is shared and accessed remotely, which increases the potential value of biometric authentication as a logical access solution. However, biometrics have proven very valuable in both types of applications.

Not every system fits neatly into the physical/logical classification. Some identification systems, especially large-scale systems, are difficult to classify because the result of a match may be to *investigate further*—there is no resultant access to data or a physical location. A biometrically enabled ATM controls access to money, a physical object, but does so by allowing a user logical access to his or her data. Even allowing for difficult-to-classify applications, the differences between logical and physical access systems are generally very pronounced: The distinction between the two is a valuable tool in understanding biometrics. Key criteria such as accuracy, response time, fallback procedures, privacy requirements, costs, and complexity of integration vary substantially when moving from logical to physical access.

How Biometric Matching Works

Because most of us are accustomed to recognizing our friends and family through their faces and voices, as well as having to *prove* who we are with passwords and keys in our day-to-day lives, you should have no problem grasping the concepts we have addressed. However, the way in which biometric technology works—the actual biometric matching functions—is more complex. In this section, we address the way biometric technologies work and the process of biometric matching.

The following is the basic process flow of biometric verification and identification (see Figure 2.2):

16   Chapter 2

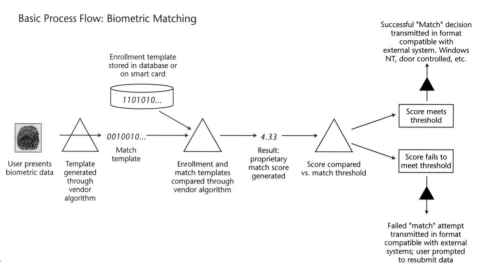

Figure 2.2 Biometric matching: process flow.

- A user initially enrolls in biometric systems by providing biometric data, which is converted into a template.
- Templates are stored in biometric systems for the purpose of subsequent comparison.
- In order to be verified or identified after enrollment, the user provides biometric data, which is converted into a template.
- The verification template is compared with one or more enrollment templates.
- The result of a comparison between biometric templates is rendered as a score or confidence level, which is compared to a threshold used for a specific technology, system, user, or transaction.
- If the score exceeds the threshold, the comparison is a match, and that result is transmitted.
- If the score does not meet the threshold, the comparison is not a match, and that result is transmitted.

Enrollment and Template Creation

The following are key terms and processes involved in enrollment and template creation:

Enrollment. The process by which a user's biometric data is initially acquired, assessed, processed, and stored in the form of a template for ongoing use in a biometric system is called *enrollment*. Subsequent verification and identification attempts are conducted against the template(s) generated during enrollment. Enrollment takes place in both 1:1 and 1:N systems, although the way a user enrolls may vary substantially from system to system. Quality enrollment is a critical factor in the long-term accuracy of biometric systems. Low-quality enrollments may lead to high error rates, including false match rate and false nonmatch rate.

Presentation. After a user provides whatever personal information is required to begin enrollment, such as name or user ID, he or she presents biometric data. *Presentation* is the process by which a user provides biometric data to an acquisition device—the hardware used to collect biometric data. Depending on the biometric system, presentation may require looking in the direction of a camera, placing a finger on a platen, or reciting a passphrase. A user may also have to remove eyeglasses or remain still for a number of seconds in order to provide biometric data. Presentation of biometric data can take as little as one second or more than one minute. The manner in which a user presents biometric data to a system is also essential to long-term performance. Users must be cognizant of the manner in which they present biometric data in order to be verified and identified successfully.

Biometric data. The biometric data users provide is an unprocessed image or recording of a characteristic. This unprocessed data is also referred to as *raw biometric data* or as a *biometric sample*. Raw biometric data cannot be used to perform biometric matches. Instead, biometric data provided by the user during enrollment and verification is used to generate biometric templates, and in almost every system is discarded thereafter. This bears repeating: Biometric systems do not store biometric data—systems use data for template creation. Table 2.1 shows data types associated with each biometric technology.

Depending on the biometric system, a user may need to present biometric data several times in order to enroll. For example, most finger-scan systems require the user to place each finger two to four times to gather sufficient data for template creation. The enrollment process may also gather data from more than one finger (or iris, or retina) to create multiple enrollment templates.

Enrollment requires the creation of an identifier such as a username or ID. This identifier is normally generated by the user or administrator during entry of personal data such as name and department. When the user returns to verify, he or she enters the identifier, then provides biometric data. Once biometric data has been acquired, biometric templates can be created by a process of feature extraction.

Table 2.1 Data Types and Associated Biometric Technologies

DATA TYPE	BIOMETRIC TECHNOLOGY
Finger-scan	Fingerprint image
Voice-scan	Voice recording
Face-scan	Facial image
Iris-scan	Iris image
Retina-scan	Retina image
Hand-scan	3-D image of top and sides of hand
Signature-scan	Image of signature and record of related dynamics measurements
Keystroke-scan	Recording of characters typed and record of related dynamics measurements

Feature extraction. The automated process of locating and encoding distinctive characteristics from biometric data in order to generate a template is called *feature extraction*. Feature extraction takes place during enrollment and verification—any time a template is created. The feature extraction process includes filtering and optimization of images and data in order to accurately locate features. For example, voice-scan technologies generally filter certain frequencies and patterns, and finger-scan technologies often thin the ridges present in a fingerprint image to the width of a single pixel. Vendors' feature extraction processes are generally patented and are always held secret. Since quality of feature extraction directly affects a system's ability to generate templates, it is extremely important to the performance of a biometric system.

Templates

The template is a defining element of biometric technology and systems, and is critical to understanding how biometrics operate. A *template* is a small file derived from the distinctive features of a user's biometric data, used to perform biometric matches. Biometric systems store and compare biometric templates, not biometric data.

There are a number of important facts about biometric templates:

➤ Most templates occupy less than 1 kilobyte, and some technologies' templates are as small as 9 bytes; template sizes also differ from vendor to vendor. Such small file sizes allow for very rapid matching, allow biometrics to be stored on devices such as tokens and smart cards, and facilitate rapid transmission and encryption.

- Templates are proprietary to each vendor and each technology. There is no common biometric template format—a template created in vendor A's system cannot be used through vendor B's technology. This is beneficial from a privacy perspective, but the lack of interoperability has deterred some would-be deployers who feared being committed to a single technology.
- Biometric data such as fingerprints and facial images cannot be reconstructed from biometric templates. Templates are not merely compressions of biometric data, but extractions of distinctive features. These features alone are not adequate to reconstruct the full biometric image or data. An analogy would be to select a string of letters from a page of text by taking the 10th letter, 20th letter, 30th letter, and so on. You would have a string of characters that, in and of themselves, had no meaning and that could not be used to rebuild the original text. However, the odds against another page of text generating that same string would be extremely slim.
- One of the most interesting facts about most biometric technologies is that unique templates are generated every time a user presents biometric data. Two immediately successive placements of a finger on a biometric device generate entirely different templates. These templates, when processed by a vendor's algorithm, are recognizable as being from the same person, but are not identical. In theory, a user could place the same finger on a biometric device for years and never generate identical templates. This is due to minute changes in positioning, distance, pressure, and various other factors that affect biometric presentation (see Figure 2.3).

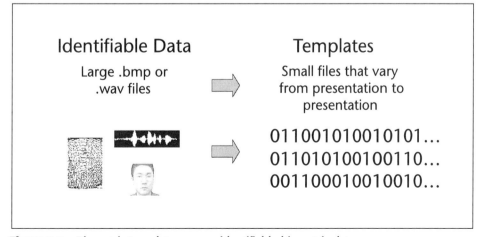

Figure 2.3 Biometric templates versus identifiable biometric data.

Depending on when they are generated, templates can be referred to as *enrollment templates* or *match templates*. Enrollment templates are created upon the user's initial interaction with a biometric system and are stored for usage in future biometric comparisons. Match templates are generated during subsequent verification attempts or identification attempts, compared to the stored template, and generally discarded immediately after the comparison. As opposed to enrollment templates, match templates are normally derived from a single sample—for example, a template derived from a single facial image can be compared to the enrollment template, which may represent an amalgam of several facial images.

Biometric Matching

The comparison of biometric templates to determine their degree of similarity or correlation is called *matching*. The process of matching biometric templates results in a *score*, which, in most systems, is compared against a *threshold*. If the score exceeds the threshold, the result is a *match*; if the score falls below the threshold, the result is a *nonmatch*.

The matching process involves the comparison of a verification template, created when the user provides biometric data, with the enrollment template(s) stored in a biometric system. In verification systems, a verification template is matched against a user's enrollment template or templates (a user may have more than one biometric template enrolled—for example, multiple fingerprints or iris patterns). In identification systems, the verification template can be matched against dozens, thousands, even millions of enrollment templates. The following are steps in involved in matching.

Scoring

Biometric match/no-match decisions are based on a *score*—a number indicating the degree of similarity or correlation resulting from the comparison of enrollment and verification templates. Biometric systems utilize proprietary algorithms to process templates and generate scores. There is no standard scale used for biometric scoring: Some biometric systems employ a scale of 1 to 100; others use a scale of -1 to 1. These scores can be carried out to several decimal points and can be logarithmic or linear. Scoring systems vary not only from technology to technology, but from vendor to vendor.

Scoring is a critical biometric concept and accounts for many of the strengths—and some of the weaknesses—of biometric systems. Traditional authentication methods such as passwords, PINs, keys, and tokens are binary, offering only a strict *yes/no* (or *no, but try again*) response. An attempt to verify via password will not succeed if it is *close*—it is either correct or incorrect. Biometric systems, by contrast, do not render absolute match/no-match decisions. Because different templates are generated each time a user interacts with a biometric system, there is no 100 percent correlation between enrollment and verification templates.

> **NOTE** Many systems return a score during enrollment, referred to as an *enrollment score* or *quality score*. This score indicates how well the feature extraction process located and encoded distinctive features from the user's biometric presentation. If the user's biometric data contains highly distinctive features or an abundance of features, there will likely be a high enrollment score. This score is often used to determine whether an enrollment attempt is successful. A low enrollment score may require that the user attempt to enroll again, perhaps by using a different spoken phrase or fingerprint. Poor enrollment scores can impact a system's ability to reliably verify or identify users.

Threshold. Once a score is generated, it is compared to the verification attempt's threshold. A threshold is a predefined number, generally chosen by a system administrator, which establishes the degree of correlation necessary for a comparison to be deemed a match. If the score resulting from template comparison exceeds the threshold, the templates are a *match* (though the templates themselves are not identical). Thresholds can vary from user to user, from transaction to transaction, and from verification attempt to verification attempt. Systems can be either highly secure or not secure at all, depending on their threshold settings. The flexibility offered by the combination of scoring and thresholds allows biometrics to be deployed in ways not possible with passwords, PINs, or tokens. For example, a system can be designed that employs a high security threshold for valuable transactions and a low security threshold for low-value transactions—the underlying comparison is transparent to the user.

Decision. The result of the comparison between the score and the threshold is a *decision*. The decisions a biometric system can make include *match*, *nonmatch*, and *inconclusive*, although varying degrees of strong matches and nonmatches are possible. Depending on the type of biometric system deployed, a match might grant access to resources, a nonmatch might limit access to resources, while inconclusive may prompt the user to provide another sample. Therefore, for most technologies, there is simply no such thing as a 100 percent match. This is not to imply that the systems are not secure—biometric systems may be able to verify identity with error rates of less than 1 in 100,000 or 1 in 1 million. However, claims of 100 percent accuracy are misleading and are not reflective of the technology's basic operation.

Biometric comparisons take place when biometric templates are processed by proprietary algorithms. These algorithms manipulate the data contained in the template in order to effect a valid comparison, accounting for variations in placement, background noise, and so on. Without the vendor algorithm, there is no way to compare biometric templates—comparing the bits that make up the templates does not indicate whether they came from the same user. The bits must be processed by the vendor as a precondition of comparison.

In most systems, enrollment and verification templates should never be identical. An identical match is an indicator that some sort of fraud is taking place, such as the resubmission of an intercepted or otherwise compromised template.

Conclusion

There are a number of biometric terms, concepts, and processes which potential deployers must fully grasp in order to make informed decisions on biometrics. Familiarity with concepts such as biometric templates and biometric matching is essential to truly understanding privacy, security, and performance in biometric systems. In particular, the controversial topic of biometric accuracy, the subject of Chapter 3, can only be fully addressed through further analysis of biometric templates and matching.

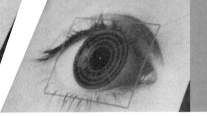

CHAPTER 3

Accuracy in Biometric Systems

For a technology that is predicated on providing strong user authentication, there is very little informative discussion of accuracy in the biometric industry. There is also very little understanding of what accuracy means in a real-world environment as opposed to a laboratory environment. With the exception of a handful of large-scale fingerprint identification systems contractually obliged to meet specific requirements, most vendor statements on accuracy bear little relevance to real-world performance of biometric systems.

Why is there such a lack of discussion and understanding of real-world performance? Because, in order to overcome the basic hurdles of developing effective matching algorithms, biometric companies have primarily concerned themselves with performance in highly controlled environments. Companies often assess their technology's accuracy—more specifically, the accuracy of their matching algorithms—by using static or artificially generated templates, images, and data, not by processing live data. When companies *do* test with actual users, the test environments do not generally replicate operation with untrained or poorly motivated subjects, as are often found in real-world deployments.

This basic algorithm-level testing is certainly a necessary step in the development of any technology, and most biometric technologies have proven their theoretical capabilities in a closed environment. However, it is necessary to evaluate and assess the capabilities of biometrics as a solution to problems in individual and institutional authentication, not as just a developmental technology. In order for companies to be comfortable deploying biometrics to employees or customers, or developing biometrically enabled product offerings, statements on biometric accuracy should reflect operation in real-world environments.

The key performance metrics in biometrics are false match rate, false nonmatch rate, and failure-to-enroll rate. No single metric indicates how well a biometric system or device performs: Analysis of all three metrics is necessary to assess the performance of a specific technology. Furthermore, because biometric system performance is affected by a variety of external factors, these performance metrics must be assessed in the context of their usage and deployment environment. Assessing accuracy without understanding the reasons behind a company's biometric deployment or the rationale for an individual's biometric usage is largely fruitless.

The following section analyzes the accuracy metrics used in biometrics and shows how these metrics can be applied to real-world applications.

False Match Rate

The first metric that comes to mind when thinking about biometric systems is *false match rate* (FMR). A biometric solution's false match rate is the probability that a user's template will be incorrectly judged to be a match for a different user's template. More bluntly, FMR describes the likelihood of an imposter beating a biometric system by being matched as someone other than him- or herself.

In a verification system, an example of a false match would be John Smith approaching a biometric system, entering Jane Doe's username or ID, presenting biometric data, and successfully matching as Jane Doe. John would then have access to any resources, physical or logical, that Jane has access to. In an identification system where there is no initial identity claim, a false match would be John Smith approaching a biometric system, presenting biometric data, and being identified from the system's database as Jane Doe.

False matches may occur because two people have similar enough biometric characteristics—a fingerprint, a voice, or a face—that the system finds a high degree of correlation between the users' templates. False match rates can be reduced by adjusting thresholds that adjust the level of correlation necessary for two templates to be judged a match. Why not simply set the threshold so that the odds of a false match are infinitesimal? Because as the false match rate is reduced, the false nonmatch rate increases. False match rate and false nonmatch rate are inversely related and must always be assessed in tandem: A system with a false match rate of 0 percent, but a false nonmatch rate of 50 percent, is secure but unusable.

In the initial development of biometric systems, proving ability to reject imposters—in other words, minimizing false match rate—was of primary importance. Accordingly, most vendor algorithms have been designed first and foremost to reject imposters; when vendors make an accuracy claim without defining their terms, they are invariably referring to false match rate. Vendors claim false match rates of 1 per 100,000, 1 per billion, even 1 in 10^{78}, although these numbers are not reflective of real-world operation. As we will see, some systems are indeed capable of extremely low false match rates, although any claimed false match rates that exceed the world's population need to be taken with a grain of salt. More important, IBG's real-world Comparative Biometric Testing has shown that some technologies and devices are much more susceptible to false matching than others.

The false match rate is often referred to as the *false acceptance rate* (FAR). The term *false acceptance rate* assumes that the result of a successful match is that the user is *accepted* into a building, an application, or a resource, which is often the case. However, in some biometric systems, the result of a false match may be that the user is locked out or denied acceptance to a resource. For example, a user may need to verify that he or she has not previously visited a facility by matching against a database of prior visitors. A false match against a prior visitor would result in *denial* of entry, not acceptance. Using FAR in this case causes confusion.

Importance of FMR

In many systems, false match rate is the most critical accuracy metric. When securing entry to a weapons facility, a bank vault, or a high-ranking system administrator's account, it is imperative that imposters be kept out. Also, as biometrics move increasingly into the public eye, solutions designed to provide high security must be perceived as relatively impervious to false match imposters. If biometrics are seen as being susceptible to imposters, they will become the target of attacks and lose credibility in the eyes of potential deployers.

An imposter break-in will certainly be a more attention-getting event than other failings of a biometric system. However, false match rate must always be balanced with false nonmatch rate and failure-to-enroll rate; these metrics have often been overlooked in the biometric industry but can be even more important to a biometric system's overall operation than false match rate.

When Are False Matches Acceptable?

A false match does not always mean that the biometric system has failed or needs threshold adjustment. Think of a video security system used by a casino to identify known card counters. Without biometrics, security personnel need to memorize the faces of every known card counter, then look at every face at every blackjack table. With biometrics, facial-scan software can scan the images provided by security cameras, locate faces, and rapidly perform 1:N identification against databases of card counters. The facial-scan system will be programmed to notify or flag any faces that come at all close to any images in the card counter database, at which point the security personnel can determine whether the two faces—one in the database, one from the live security camera—are a match. Even if 9 out of 10 matches are incorrect, the facial-scan system is performing a valuable function by alerting staff to possible matches and reducing overall manual effort.

Single False Match Rate versus System False Match Rate

As commonly used in the biometric industry, the false match rate states the likelihood of a false match for a *single comparison* of two biometric templates. This can be thought of as a single false match rate. In real-world applications, however, a person trying to break into a biometric system may be able to attempt more than one match—he or she may be able to try a series of matches. As a result, the actual odds of an imposter break-in are much higher. The likelihood of an imposter break-in for a given system is the system false match rate.

To illustrate, let's take a finger-scan system with a stated false match rate of 1/10,000. In this system, the likelihood of templates from two different people matching during a single match attempt is 1/10,000. Let's also assume that the system is not designed to lock a user out after a certain number of attempts. John Smith enters Jane Doe's username, places his index finger, and is rejected. John Smith can then try to break in with all 10 fingers—each break-in attempt has a false match rate of 1/10,000, but the overall system false match rate is immediately higher. John doesn't break in against Jane, but happens to have Kris's username as well. John can attempt to break in 10 more times, further increasing the false match rate. If the verification environment is unsupervised, this might continue unabated until the likelihood of a false match becomes dangerously high.

As this makes clear, *the most important false match metric in real-world deployments is the system false match rate.* Real-world users and deployers of biometric technology are concerned with how well biometrics will solve their identification and authentication problems. The system false match is a much more useful expression of problem-solving capability than the single false match rate alone. Determining system false match rate requires knowledge of the single false match rate, the number of verification or identification attempts the user is allowed, the availability of usernames an imposter has access to, and other factors that can be determined only by looking at system process flow. Although ensuring a low system false match rate is a more complicated task than simply assessing a technology's false match rate, it is a requirement of deploying secure biometric systems.

False Match Rate in Large-Scale Identification Systems

The expectations regarding FMR are often different for large-scale identification systems. Because over a billion matches might take place on a given day, large-scale identification requires much lower false match rates than would commonly be the case in verification systems, lest system administrators be flooded with thousands of false matches.

In particular, negative identification systems—those designed to ensure that an individual's biometric data is not present more than once in a database—require low false match rates, as it is assumed that most users will not be attempting to commit fraud. Since the very presence of the system serves as a deterrent, it is not worthwhile to flag and investigate every near match. Instead, only very strong potential matches need be flagged and investigated. Therefore, the false match rate must be kept as low as possible.

False Nonmatch Rate

A biometric solution's *false nonmatch rate* (FNMR) is the probability that a user's template will be incorrectly judged to *not* match his or her enrollment template. In most cases, a false nonmatch means that an authorized user is locked out of a system, incorrectly denied access to a facility or resource.

In a verification system, an example of a false nonmatch would be John Smith approaching a biometric system, entering his username or ID, presenting biometric data, and failing to match. In an identification system, a false nonmatch would be John Smith approaching a biometric system in which he had previously registered, presenting biometric data, and not being located in the database.

A solution's false nonmatch rate is often referred to as *false rejection rate* (FRR). This terminology is appropriate when the result of a false nonmatch is a user being denied access to a facility or resource, but this is not always the case. For example, a system may deny access to users who are found in a particular database. If a previously enrolled user presents his or her biometric data and is not matched in the database, he or she has fraudulently gained access to a resource as a result of a false nonmatch.

False nonmatches occur because there is not a sufficiently strong correlation between a user's verification and enrollment templates. This can be attributed to the following (see also Table 3.1 on pages 30 and 31):

➢ Changes in a user's biometric data
➢ Changes in how a user presents biometric data
➢ Changes in the environment in which data is presented

Because each of these is a fairly common occurrence, biometric systems are much more susceptible to false nonmatches than they are to false matches. Well-designed systems can reduce the impact of changes in biometric data, but most systems begin to see at least slight increases in false rejection over time.

Importance of FNMR

In most biometric applications, particularly those involving employees, the vast majority of verification attempts will be genuine. As a result, high false nonmatch rates can be as damaging to an enterprise as high false match rates. When users are falsely nonmatched and denied access to resources, the result is lost productivity, frustrated users, and an increased burden on help desk or support personnel. Biometric vendors have traditionally focused on limiting false match rates, but a disproportionate focus on FMR reduction can lead to unacceptably high false nonmatch rates.

Changes in a User's Biometric Data

Many biometric technologies are based on characteristics that can change between enrollment and verification, especially as a substantial amount of time elapses. Voice-scan systems, for example, do not perform as well when users have sore throats; facial-scan systems perform poorly if a user has experienced drastic changes in facial hair or weight. Even more stable characteristics such as fingerprints can change over time, as scars, aging, and general wear and tear lower the quality of the data. These temporal changes are one of the more imposing challenges that biometric systems face, but the issue of accuracy over time is very rarely broached in the biometric industry.

Technologies somewhat susceptible to changes in underlying biometric data include facial-scan and hand-scan; eye-based biometrics, iris-scan, and retina-scan are less susceptible. Long-term robustness of finger-scan solutions varies too widely from vendor to vendor to make a technology-wide assessment.

Changes in User Presentation

Even if the biometric data has not changed in any way, the manner in which a user presents data to a biometric system can lead to substantial false rejection rates. Changes between enrollment and verification presentations are a common cause of false nonmatching—placing a finger at a different angle, positioning one's self more closely to a camera, speaking at a different volume. Although biometric systems can overcome *some* variance in user presentation between enrollment and verification, after a certain point, the users' data will not match.

These changes are not necessarily attributable to user carelessness. Some biometric systems are designed to facilitate proper presentation, whereas others are very difficult for users to interact with consistently. To ensure more consistent user presentation, most biometric systems provide feedback to users during the presentation process. Finger-scan systems may advise users to move a finger up or down; voice-scan systems may advise users to speak more softly or loudly. In addition, if not trained properly, users may not be cognizant of the manner in which they enrolled, meaning that they will be unable to replicate the process. Situations in which verification is monitored can reduce the likelihood of false nonmatches, as users can be instructed on the quality of their presentation.

Behavioral biometrics, in particular, are subject to changes in user presentation. Signature-scan, facial-scan, and voice-scan require special attentiveness in order for the verification attempts to match the enrollment presentation. Eye technologies are somewhat susceptible to changes in user presentation, whereas finger-scan devices have been designed to minimize user presentation errors.

Changes in Environment

Assuming that the biometric data has not changed and the user presents his or her biometric data correctly, false nonmatches can still occur as the result of changes in authentication environment. Variations in background lighting and composition, noise, and even temperature can lead to increased false nonmatch rates. Enrolling on a landline telephone, for example, and attempting to verify via mobile phone is likely to lead to high FNMR. Facial-scan cannot accommodate significant variation in lighting between enrollment and verification: Attempting to verify in a backlit environment against an enrollment that took place under frontal lighting will result in high FNMR. A well-designed facial-scan system, one that ensures that enrollment and verification take place under similar conditions, can greatly reduce this problem. Large-scale identification systems, for example, ensure that environmental variance between enrollment and verification is limited.

Table 3.1 Factors Affecting False Nonmatch Rates

TECHNOLOGY	SUSCEPTIBILITY TO CHANGES IN BIOMETRIC DATA	SUSCEPTIBILITY TO CHANGES IN USER PRESENTATION	SUSCEPTIBILITY TO ENVIRONMENTAL CHANGES
Finger-scan	*Low to moderate.* Some user groups are prone to changes in fingerprint quality; for most users, the fingerprint remains stable and usable.	*Moderate.* Variation in angle, pressure, and position can lead to high FNMR.	*Low.* Some solutions are susceptible to light and temperature changes, but this is rare.
Facial-scan	*Moderate to high.* Changes in facial hair and hairstyle, as well as drastic changes in weight, can increase FNMR.	*Low to moderate.* In a controlled environment, user presentation should not vary. Otherwise, user is responsible for quality of presentation. Removal of glasses and hats can also increase FNMR.	*Moderate to high.* Changes in intensity and angle of lighting, as well as background composition, can be highly problematic for facial-scan.
Voice-scan	*Low to moderate.* Illnesses can affect the voice to the point where FNMR increases.	*Moderate to high.* Volume of speech, inflection, and duration must be consistent to ensure verification.	*Moderate to high.* Background noise and quality of acquisition device (microphone, telephone) can increase FNMR.
Iris-scan	*Low.* The iris is highly stable and, short of eye trauma, iris patterns do not change.	*Moderate to high.* Users must be positioned correctly in order for high-quality image to be captured.	*Low.* Lighting does not affect the technology, as it uses infrared illumination.
Signature-scan	*Moderate to high.* Data and presentation cannot be effectively separated. For some users, the signature is highly susceptible to variation, especially over time. Signing in different position (sitting vs. standing) may impact performance.		*Low.* Technology largely unaffected by environment.
Hand-scan	*Moderate.* Injuries, swelling, water retention can increase likelihood of false nonmatching.	*Low to moderate.* Design of technology facilitates correct placement, but placement itself can be challenging for some users.	*Low.* Technology largely unaffected by environment.

TECHNOLOGY	SUSCEPTIBILITY TO CHANGES IN BIOMETRIC DATA	SUSCEPTIBILITY TO CHANGES IN USER PRESENTATION	SUSCEPTIBILITY TO ENVIRONMENTAL CHANGES
Retina-scan	*Low.* Aside from dislocation due to age-related eye diseases, retinal patterns are stable.	*High.* User presentation is extremely challenging, leading to atypically high FNMR.	*Low.* Technology largely unaffected by environment.
Keystroke-scan	*Moderate to high.* Data and presentation cannot be effectively separated. Typing patterns easily change, especially for poor typists. Effect of different keyboard types (click vs. nonclick, ergonomic keyboards) unknown.		*Low.* Technology largely unaffected by environment.

Real-World False Nonmatch Rates

Given that so many factors can contribute to FNMR, what sort of accuracy rates are encountered in production systems? Vendors do not often discuss FNMR by itself, as it has been seen as a secondary measure in comparison to false match rate. Those who do will cite false nonmatch rates in the vicinity of 1 percent, 0.1 percent, and 0.01 percent. Real-world testing has shown that these rates are, in many cases, wildly optimistic. Biometric systems from well-funded, established companies have shown false nonmatch rates in the high single digits, even reaching double-digit percentages, *immediately after enrollment*. Amazingly, these results are not derived from a single presentation of biometric data, but are derived from multiple verification attempts. For many technologies, the results are even worse after time elapses. Increases in false nonmatches of 100 percent, 200 percent, even 400 percent are not unheard of. These results suggest that the very low false nonmatch rates claimed by vendors are not based on interaction with live subjects.

At the same time, other technologies tested side by side with these poor performers have showed false nonmatch rates at or near 0 percent, even after weeks have elapsed. Clearly, there is no all-encompassing rule regarding biometric systems' resistance to false nonmatches.

Single FNMR versus System FNMR

As was the case when discussing false match rates, generating false nonmatch rates from a single presentation or verification attempt is inconsistent with real-world system usage. In operational deployments, it is rare that users have only one chance to present data to a biometric system. Instead, users are generally allowed to attempt verification a handful of times until an account or login name is locked. As a result, the single false nonmatch rate, which represents the probability of a single user attempt resulting in a false nonmatch, does not reflect real-world usage. Instead, the *system false nonmatch rate* is a better indicator of how a system will perform in a real-world environment.

As an illustration, consider a system that has a single false nonmatch rate of 1 percent, meaning that 1 percent of verification attempts are incorrectly judged to not match. For most deployments, this would be an unacceptably high rate. However, users normally have between three and five attempts to verify in logical and physical access situations, allowing for improved presentation and thereby reducing the effective FNMR. In addition, most finger-scan systems strongly recommend enrolling more than one finger to reduce FNMR. By attempting to verify with a second or third finger and allowing multiple verification attempts, the system false nonmatch rate becomes much less than 1 percent.

In contrast to system false match rate, which recasts the real-world FMR of biometric systems in a fairly harsh light, utilizing the system false nonmatch rate is beneficial to biometric systems. Because of the multiple factors that can contribute to false nonmatching, some of which cannot be controlled or accounted for by biometric vendors, it is reasonable to assume that a user may need to attempt verification more than once in order to be verified. Though the user may be slightly inconvenienced by having to place a backup finger or recite a passphrase more than once, the inconvenience of being denied access would likely be more substantial.

False Nonmatch Rates in Large-Scale Identification Systems

Whereas a 5 percent false nonmatch rate in a logical access verification system would be cataclysmic, large-scale identification systems can operate very effectively with a 5 or 10 percent FNMR. In a fraud detection and deterrence system, identifying 90 to 95 percent of fraudulent attempts represents a dramatic improvement over the capabilities of nonbiometric systems. Given that the existence of the system will deter some percentage of users from even attempting to commit fraud, 5 to 10 percent may be more than acceptable.

Furthermore, many of the issues that affect verification systems, such as environmental factors and user presentation, can be accounted for more effectively in large-scale systems; enrollment and verification are normally strictly controlled events. Users are unlikely to circumvent the system due to poor placement or background variations. Therefore, the 5 to 10 percent FNMR can be tuned by increasing or decreasing the strictness of the matching algorithm, because decisions are being based on consistent presentations.

Failure-to-Enroll (FTE) Rate

Traditionally, false match rate and false nonmatch rate have been the basic criteria by which biometric accuracy was assessed. This is consistent with the industry's focus on testing static images and data, using compliant and knowledgeable test subjects, and focusing strictly on matching algorithm capabilities. Vendors have paid less attention to the third critical accuracy metric: failure-to-enroll rate.

A system's *failure-to-enroll* (FTE) rate represents the probability that a given user will be unable to enroll in a biometric system. FTEs occur when users have insufficiently distinctive or replicable biometric data or when the design of the biometric solution is such that providing consistent data is difficult. High failure-to-enroll rates can be particularly problematic for a biometric system, as users unable to enroll must verify through another biometric technology or authentication method.

In order to define a failure-to-enroll, one must first define enrollment. As it happens, the process differs significantly from technology to technology and from device to device. Both physiological and behavioral biometrics are subject to failures-to-enroll. Finger-scan systems may require between one and six high-quality presentations of a single fingerprint, and may require that a user enroll two fingers for an enrollment to be complete. Iris-scan systems can require between one and four images captured per iris. Voice-scan systems may require that a passphrase be recited three times or that a string of numbers be repeated for 30 to 40 seconds. Facial-scan enrollment may require the capture of a handful of images, may be based on duration of image capture, or may take place through a static image. The only common element of these varied enrollment processes is the result: A user's information is eventually stored in some type of database or file for future comparisons.

More so than for false match and nonmatch rates, a system's FTE rate is dependent on system design, training, and ergonomics—not necessarily on the underlying biometric data. Improvements in system design and enrollment process can reduce a system's FTE rate from 10 percent to 1 percent, without any changes to the core biometric processes. For example, asking a user to shorten a spoken or signed passphrase or enrolling alternate fingerprints can dramatically reduce FTEs. Ensuring proper placement and interaction with acquisition devices can also drastically reduce FTE.

Conversely, even the best biometric algorithms are of little use when enrollment processes make it difficult for users to provide data. Consider retina-scan: In theory, retina-scan is a very appealing biometric, as it is based on highly distinctive and stable characteristics and cannot be utilized without user consent. However, the enrollment process is quite burdensome, requiring users to keep their eye and head perfectly still for several seconds while focusing on a small light. Because of this, a high percentage of users are unable to enroll in retina-scan, reducing the areas in which it can be effectively deployed. (See Table 3.2.)

Table 3.2 Failure-to-Enroll by Technology

TECHNOLOGY	CAUSES OF FTE
Finger-scan	Low-quality fingerprints (scarred, faint, no breaks in ridges)
	Inconsistent submission (angle and pressure of placement)
	Difficult submission process (sliding, swiping finger)
Facial-scan	Glasses, hats, headwear
	Poor lighting
	Background composition
Voice-scan	Background noise
	Quality of acquisition device
	Inconsistent vocal samples (volume, duration, inflection)
Iris-scan	Difficult submission process (lining up eye with guides, gauging correct distance)
	Glasses (cause reflection)
Signature-scan	Inconsistent signatures
Hand-scan	Inability to position hand according to guide pegs
Retina-scan	Difficult submission process (must remain still for several seconds for each acquisition, multiple acquisitions required)
Keystroke-scan	Inconsistent typing patterns

Importance of FTE Rates

The impact of FTE differs for individual and institutional users. For individuals unable to enroll in a biometric system or device designed for personal use, inability to enroll may be frustrating, but they will still have recourse to standard authentication. For institutions offering biometric authentication to customers, FTE becomes a customer service issue and will lead to disgruntled customers. However, even this is not insurmountable—it is unlikely that an institution will penalize nonbiometric users, so recourse to alternate authentication is still an option.

FTE can be a major problem in internal, employee-facing deployments. In this environment, high failure-to-enroll rates are directly linked to increased security risks and increased system costs. Consider a biometric network authentication system. Deploying a technology with a 2 percent FTE rate to 1,000 users means that approximately 20 users will be unable to verify biometrically. These FTE users will need to be authenticated in some fashion. If they revert to password authentication, then there are 20 accounts on the network that are just as susceptible to compromise as before biometrics were deployed. Furthermore, the remnants of a password system are still in place, meaning that an infrastructure must be in place for password maintenance and changes.

Relation of Failure-to-Enroll to False Nonmatch Rate

False match rates and false nonmatch rates are inversely related: Decrease one rate and the other rate increases, and vice versa. Interestingly, adjusting system settings to reduce FTE rates can also affect false nonmatch rates. By reducing the number of distinctive or replicable features required for enrollment, FTE rates can be lowered—marginal images and data can then be enrolled. The result is that users will be more likely to be falsely nonmatched in subsequent verification attempts, because it is more difficult for a system to find data elements in common with marginal enrollments.

FTE Across Different Populations

Certain ethnic and demographic populations are more prone to high FTE rates than others, a problem that can reduce the deployability of biometrics in specific environments. This problem has been studied most thoroughly in finger-scan, but evidence suggests that other biometric technologies may suffer from similarly inconsistent performance across demographic lines.

> ### Real-World Performance Testing
>
> In order to determine which of today's biometric technologies and solutions perform best in terms of false match rates, false nonmatch rates, and failure-to-enroll rates, real-world testing is an indispensable tool. Deploying biometrics without an understanding of how well they perform in operational environments can lead to failed deployments and vulnerable systems.
>
> Since 1998, International Biometric Group has conducted comparative testing of leading biometric systems to determine how well they perform under controlled conditions. Each test cycle involves hundreds of individuals unfamiliar with biometric technology enrolling and verifying in 10 off-the-shelf biometric systems.
>
> In order to eliminate variables that might skew results, the testing follows a consistent set of protocols for each technology tested, including finger-scan, facial-scan, voice-scan, iris-scan, signature-scan, and keystroke-scan. Users are allowed a fixed number of enrollment attempts, *false verification* attempts (also known as *imposter break-in* attempts), and *true verification* attempts. In addition, users return after six weeks in order to verify against their original enrollments.
>
> The results of recent test cycles show that today's biometric technologies, on the whole, are much more capable of strong performance than technologies from 1998 through 1999. At first it was a challenge just to keep all 10 biometric systems operating for the months required to finish testing; many technologies have error rates well above 50 percent. In recent cycles, systems have shown themselves to be much more stable, and a greater percentage of solutions offer accuracy well above 95 percent. Testing also shows that performance can vary drastically within technologies—some finger-scan solutions, for example, had next to no errors during testing, while others rejected nearly one-third of enrolled users. Most interestingly, the testing shows that over time, many biometric systems are prone to incorrectly rejecting a substantial percentage of users. Verifying a user immediately after enrollment is not highly challenging to biometric systems. However, after six weeks, testing shows that some systems' error rates increase tenfold.

In comparative testing of finger-scan technology, three types of users—elderly, construction workers/artisans, and those of Pacific Rim/Asian descent—are more prone to FTE than control groups. Elderly users often have very faint fingerprints and may have poorer circulation than younger users. Construction workers and artisans are more likely to have highly worn fingerprints, to the point where ridges are nearly nonexistent. Users of Pacific Rim/Asian descent may have faint fingerprint ridges—especially female users. Since robust finger-scan solutions generally have FTE rates in the vicinity of 1 percent, a 100 percent increase in FTE for a specific population would result in a 2 percent FTE rate—not excessively high, but still warranting attention.

Testing of facial-scan solutions indicates that the technology may not be as adept at enrolling very dark-skinned users. The increased FTE rate is not attributable to the lack of distinctive features, of course, but to the quality of the images provided to the facial-scan system by video cameras optimized for lighter-skinned users. Though not conformed in testing, one of the potential limitations of iris-scan technology is the ability to locate distinctive features in very dark irises. Iris-scan technology is based on 8-bit grayscale image capture, which allows for 256 shades of gray; features from very dark irises may be clustered at one end of the spectrum.

Single FTE Rates versus System FTE Rates

There is much less discussion in the biometric industry of FTE than of other accuracy metrics. Therefore, there is no consensus on what constitutes a failure-to-enroll. The strictest definition of a single failure-to-enroll rate would be the likelihood that a user is unable to enroll after one full enrollment sequence. However, in real-world applications, users have more than one chance to enroll. A user may simply be prompted to attempt enrollment a second time, or may be routed to a training and presentation sequence. A user may be prompted to place a second finger or to choose a longer or shorter passphrase. Biometric systems are normally quite flexible in allowing for multiple enrollment attempts.

The *system failure-to-enroll rate* represents the percentage of users who are deemed FTEs after a reasonable number of attempts to enroll in a biometric system. This number will always be lower than the *single failure-to-enroll rate* because many users require one enrollment attempt to become acclimated to interacting with a given biometric system.

The point at which a user becomes an FTE is dependent on the application. In a network authentication system, a user may be allowed to attempt enrollment several times, because enrollment takes place at his or her desktop. On the other hand, a deployer may require that enrollment be kept very brief—two full attempts, for example—so as not to reduce productivity. Users unable to immediately enroll in a physical access system may be asked to reattempt enrollment at a later time to avoid generating long lines. Somewhere in every application, a decision point is reached that concludes that a given user is, indeed, a failure-to-enroll. Keep in mind that a user who requires 10 attempts to enroll is very likely to encounter high false nonmatch rates, as he or she is clearly unable to provide consistent biometric data. There is a point of diminishing returns when making multiple attempts to enroll a user.

FTE in Large-Scale Identification Systems

In large-scale identification systems, the result of a failure-to-enroll is that a duplicitous user may be able to register for a service or program multiple times. Although certainly not desirable, this problem is not crippling to large-scale identification systems. Even if a small percentage of users cannot enroll, enough users will be successfully enrolled to detect and deter a significant portion of fraud. Furthermore, the user is not necessarily informed of the quality of his or her enrollment attempts. A user may be unaware that he or she is a failure-to-enroll in a large-scale identification system and would have no reason to think that a fraud attempt might be successful.

Derived Metrics

In addition to the aforementioned metrics, there are two metrics used to reflect the overall accuracy capabilities of a biometric technology. These derived metrics are generated from analysis and comparison of FMR, FNMR, and FTE.

Equal Error Rate (EER)

The *equal error rate* (EER), also referred to as the *crossover rate*, is the rate at which the FNMR is equal to the FMR. That is, it represents the accuracy level at which likelihood of a false match is the same as the likelihood of a false nonmatch. Equal error rate is commonly used as a representation of overall system accuracy, as it is a general indicator of a system's resistance to break-ins and ability to match templates from authorized users. However, if used to make determinations on a technology's capabilities, EER can be both irrelevant and misleading.

There are very few applications in which the expectations for security (the ability to reject unauthorized users) identically match expectations for convenience (the ability to match authorized users). In most applications, the ability to reject imposters takes precedence, as it is assumed that what is being protected is of some value. It is very rare that a system would ever be deployed using the EER as a guideline. Furthermore, EER is generated from single false match and single false nonmatch rates, neither of which is reflective of system operation in the real world. Because of this, EER can be highly misleading. System false match and nonmatch rates are a much more reliable indicator of real-world capabilities.

Most important, EER does not incorporate failure-to-enroll rates. A system might have an extremely low EER but an FTE rate of 15 percent. Looking solely at EER would give a false sense of a system's capability and deployability in real-world environments.

Ability-to-Verify (ATV) Rate

A more valuable derived metric is the *ability-to-verify (ATV)* rate. ATV is a combination of the failure-to-enroll and false nonmatch rates, and indicates the overall percentage of users who will be capable of authenticating on a daily basis. ATV is rendered as follows:

$$ATV = (1 - FTE)(1 - FNMR)$$

This metric can be thought of as representing the group of users who cannot enroll (FTE) along with users falsely rejected by the system (FNMR). No system has a 100 percent ATV rate, but, in general, a high ATV rate will make for a more effective system. When balanced with an acceptable FMR, ATV can be extremely useful because it is has an impact on three key aspects of biometric deployments: cost, security, and convenience.

Cost. One of the most expensive aspects of a biometric system is the costs involved with exception processing. Any user unable to be processed by the biometric needs to be processed by a fallback procedure, meaning that dual systems must be maintained. Whether an alternate biometric, a password, or a live verification, there is a need for a separate enabling and support infrastructure.

Security. A low ATV means that a substantial percentage of users are not being verified by your system. The security provided by a system that can verify only 90 percent of its users may be acceptable for some deployments, but can be problematic in many others.

Convenience. A low ATV may be a reflection of a difficult-to-use system. In situations in which user convenience is paramount, adjustments to enrollment and verification settings may be required to maximize the ATV rate.

Recall that comparison of biometric templates does not result in 100 percent matches, but instead results in some degree of correlation. A score is generated after template comparison; this score is then compared to a threshold, and a decision is rendered. Biometric systems return degrees of certainty, not 100 percent certainty.

The security, convenience, and fraud reduction that biometrics are expected to provide are based on the assumption that biometrics work—that they verify and identify users correctly and can enroll a high percentage of users. Just as defining the best biometric is fruitless without understanding the application, assessing biometric accuracy requires that the application be defined.

Conclusion: Biometric Technologies from an Accuracy Perspective

Biometric systems can make mistakes and are not always able to enroll every user. To determine whether a particular biometric technology can meet performance requirements, institutions must assess all three accuracy metrics; assessing anything less than all three metrics is not only of reduced value, but can be highly misleading. Using only selected performance metrics will generate a false sense of the system's actual capabilities. Furthermore, the interdependency of the three metrics is a critical element of biometric accuracy; improvements in one area may lead to problems in another.

Most important, assessing system rates as opposed to single rates gives the strongest indication of a biometric technology's real-world suitability for a given application. System rates are much more difficult to derive, as they are contingent on combining system design with single false match rates, but they are essential to understanding the security, convenience, and cost savings that a biometric system provides.

As we move on to address biometric technologies, it is essential to remember that biometric accuracy impacts system security, convenience, and cost. Deploying biometrics intelligently requires that real-world accuracy be taken into account in employee, customer, and citizen-facing applications.

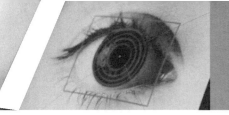

PART TWO

Leading Biometric Technologies: What You Need to Know

Perhaps the most common question posed to those in the biometric industry is, *What is the best biometric?* Unfortunately, this question can only be answered with another series of questions. Is *best* defined as most capable of rejecting false attempts? Most capable of matching authorized users? Most resistant to spoofing? Least expensive to deploy? Most privacy-protective? Most likely to accommodate every potential user, regardless of age, gender, and background? Easiest to use? Best for what application, for what demographic, in place of what authentication technology, with what level of support? Unless these questions are answered, there is no way to determine what is the best biometric technology.

The leading biometric technologies—including finger-scan, facial-scan, voice-scan, hand-scan, iris-scan, signature-scan, retina-scan, keystroke-scan, and Automated Fingerprint Identification Systems (AFISs)—each have their strengths and weaknesses, and are each well-suited for particular applications. There is no single best biometric technology, nor is it likely that any single technology will come to dominate in every area of the biometric industry. Instead, the *requirements of a specific application* determine which, if any, is the best biometric. While comparing the capabilities of finger-scan, facial-scan, and other leading technologies as a solution for a specific application—such as desktop 1:1 verification—is very valuable, attempting to compare biometric technologies without an application is largely meaningless.

The chapters in this section tell potential biometric deployers what they need to know about leading technologies: their components, their operations, the applications for which they are successfully deployed, and their strengths and weaknesses.

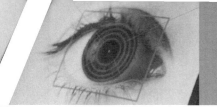

CHAPTER 4

Finger-Scan

Finger-scan technology utilizes the distinctive features of the fingerprint to identify or verify the identity of individuals. Finger-scan technology is the most commonly deployed biometric technology, used in a broad range of physical access and logical access applications. Dozens of finger-scan vendors compete in this marketplace, offering hardware devices, software packages, enterprise solutions, and standalone solutions. (see Figure 4.1.)

Finger-scan's strengths include the following:

➢ It is a mature and proven core technology, capable of high levels of accuracy.

➢ It can be deployed in a range of environments.

➢ It employs ergonomic, easy-to-use devices.

➢ The ability to enroll multiple fingers can increase system accuracy and flexibility.

Finger-scan has a handful of weaknesses that may impact its effectiveness for certain applications:

➢ Most devices are unable to enroll some small percentage of users.

➢ Performance can deteriorate over time.

➢ It is associated with forensic applications.

Precise Biometrics SC-100

Sony FIU-710

Bioscrypt Veriprox

Figure 4.1 Finger-scan devices.

Components

Finger-scan systems comprise image acquisition hardware, image processing components, template generation and matching components, and storage components. These components can be located within a single peripheral or standalone device, or may be spread among a peripheral device, a local PC, and a central server.

The surface on which the finger is placed is called a *platen,* also referred to simply as a *scanner.* Platens can be made of various materials, including glass, plastic, silicon, and polymer. Proprietary coatings are used to prevent damage to the platen itself—a scratched platen reduces a device's ability to acquire high-quality fingerprint images. Depending on the type of finger-scan technology used, areas of contact between the fingerprint and the platen are measured through chip-based cameras, through ultrasonic imaging, or through changes in capacitive fields generated by the finger. These measurements are converted into digital code, at which point they can be processed by the finger-scan system.

A platen is one piece of a finger-scan *module,* the basic building block of a peripheral or standalone finger-scan device. A module normally contains a platen attached to a small printed circuit board, along with a standard connector that allows digitized information to be transmitted to the peripheral or standalone device. Many of today's finger-scan modules are capable of performing all system functions—image acquisition, image processing, template generation, template matching, and template storage—within a very small package. These modules simply send match/no-match decisions to an external application or system (see Figure 4.2).

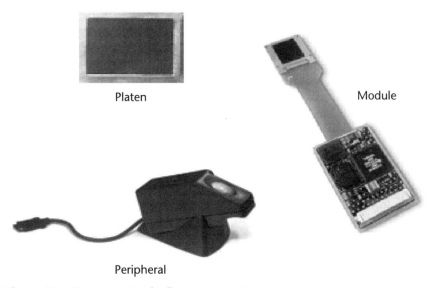

Figure 4.2 Components of a finger-scan system.

Modules can be built into PC peripheral devices, embedded in keyboards, laptops, handheld devices, or PCMCIA cards; integrated into door control devices; or built into standalone terminals and readers. Eventually they may be commonly found in mobile phones, on the surface of smart cards, and in motorized vehicles. Depending on cost and system design requirements, peripherals designed for PC access can perform all functions internally. However, in most finger-scan systems, template generation and matching, being more processor-intensive activities, take place on a local PC or central server. Standalone and physical access devices are more likely to have built-in processing power, such that all of the biometric processes take place within the device itself (see Figure 4.3).

The results of a finger-scan match are transmitted from the biometric system in the proper format for authentication to logical or physical access systems. Finger-scan systems may be tightly integrated with these external systems (applications, operating systems, smart card authentication components, electronic door locks, mobile phones) or may simply pass a password to the application. This integration piece is an extremely important part of any biometric system and is especially so for finger-scan technology due to the wide range of environments in which finger-scan authentication is used to grant or limit access to data, resources, or a physical area.

How Finger-Scan Technology Works

There are five stages involved in finger-scan verification and identification: fingerprint image acquisition, image processing, location of distinctive characteristics, template creation, and template matching. Though each vendor's process may differ slightly from the sequence described, especially in feature extraction, the basic process is similar.

Image Acquisition

The first challenge facing a finger-scan system is to acquire a high-quality image of the fingerprint. Image quality is measured in dots per inch (DPI)—more dots per inch means a higher-resolution image. Today's finger-scan peripherals can acquire images of 500 DPI, the standard for forensic-quality fingerprinting. The lowest DPI generally found in the market is in the 300- to 350-DPI range (see Figure 4.4).

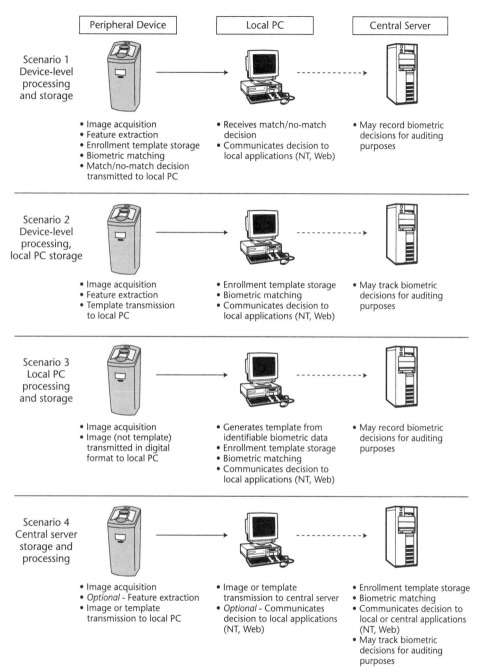

Figure 4.3 Schematic—data storage and processing in finger-scan systems.

Figure 4.4 Typical finger-scan images.

Image acquisition is a major challenge for finger-scan developers, because fingerprint quality can vary substantially from person to person and from finger to finger. Some populations are more likely than others to have faint or difficult-to-acquire fingerprints, whether due to wear and tear or physiological traits. In addition, environmental factors can impact image acquisition. In very cold weather, the oils normally found on a fingerprint (which make for better imaging) dry up, such that fingerprints can appear faint. Users may need to press more firmly or even rub the finger into their opposite palm to ensure that a quality image is acquired.

For a finger-scan image to act as an effective enrollment, the center of the fingerprint must be placed on the platen. Many users unfamiliar with the technology will place their finger at an angle, such that only the upper portion of the fingerprint appears. This results in fewer distinctive features being located during enrollment and verification, reducing the likelihood of successful operation.

An additional factor in image acquisition that can affect a system's accuracy and performance is the size of the platen. Over time, finger-scan vendors have developed smaller and smaller platens in order to manufacture smaller devices and to reduce costs. However, there may be a point of diminishing returns in terms of minimizing platen size. Very small platens acquire a smaller portion of the fingerprint, meaning that less data is available to create and match templates. Users with large fingers may also find it difficult to present their fingerprint in a consistent fashion, leading to false rejections.

Image Processing

Once a high-quality image is acquired, it must be converted to a usable format. Image processing subroutines eliminate gray areas from the image by converting the fingerprint image's gray pixels to white and black, depending on their pitch. What results is a series of thick black ridges (the raised part of the fingerprint) contrasted to white valleys. The ridges are then thinned from approximately 5 to 8 pixels in width down to a single pixel, for precise location of features (see Figure 4.5).

Location of Distinctive Characteristics

There is a great deal of distinctive information on the average fingerprint—enough to enable large-scale searches using only one or two fingerprints. This information is fairly stable throughout one's life and differs from fingerprint to fingerprint, even for identical twins.

The fingerprint comprises ridges and valleys that form distinctive patterns, such as swirls, loops, and arches. Most fingerprints also have a core, a central point around which swirls, loops, or arches are curved. Deltas are points, normally at the lower left or right corner of the fingerprint, around which ridges are centered in a triangular shape.

Fingerprint ridges and valleys are characterized by discontinuities and irregularities known as *minutiae*—these are the distinctive features on which most finger-scan technologies are based. There are many types of minutiae, the most common being ridge endings (the point at which a ridge ends) and bifurcations (the point at which one ridge divides into two). Depending on the size of the platen and the quality of the vendor algorithm, a typical finger-scan image may produce between 15 and 50 minutiae—larger platens will acquire more of the fingerprint image, meaning that a greater number of minutiae can be located.

Raw Image Ridge Isolation Thinned Ridges

Figure 4.5 Steps involved in image processing.

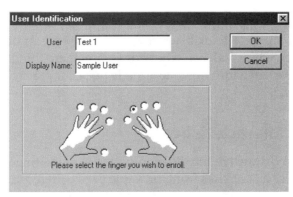

Figure 4.6 Finger-scan enrollment prompts.

Template Creation

Vendors utilize proprietary algorithms to map fingerprint minutiae. Information used when mapping minutiae can include the location and angle of a minutia point, the type and quality of minutiae, and the distance and position of minutiae relative to the core. A user normally must place his or her enrollment fingerprint more than once during enrollment, so that the system can locate the most consistently generated minutiae (see Figure 4.6).

Finger-scan images will normally have distortions and *false minutiae* that must be filtered out before template creation. For example, anomalies caused by scars, sweat, or dirt can appear as minutiae. Vendor algorithms scan images and eliminate features that simply seem to be in the wrong place, such as adjacent minutiae or a ridge crossing perpendicular to a series of other ridges. A large percentage of false minutiae are discarded in this process, ensuring that the template generated for enrollment or verification is an accurate reflection of the biometric data (see Figure 4.7).

Template Matching

Finger-scan templates can range in size from approximately 200 bytes to over 1,000 bytes—a very small amount of data by any measure. These templates cannot be manually read as anything resembling a fingerprint, and simply

Figure 4.7 Finger-scan minutiae.

performing a bit-to-bit comparison of two finger-scan templates will not determine whether they are from the same person. Instead, vendor algorithms are required to process templates and to determine the correlation between the two.

Comparing enrollment and verification templates does not result in an exact match. The position of a minutia point may change by a few pixels, some minutiae will differ from the enrollment template, and false minutiae may be seen as real. Also, the fingerprint will inevitably be placed at a slightly different angle. However, matching algorithms can account for these variations and allow for effective comparison of templates in which much of the underlying data may have changed.

There is no minimum number of minutiae necessary for two finger-scan templates to match. In some cases, the system may need to locate only a handful of minutiae in common to decide that two templates are a match. Higher system thresholds will require that a higher percentage of the minutiae points match and can require more careful placement during verification. If a finger-scan is deployed for 1:few identification against modest databases as opposed to 1:1 verification, these thresholds will likely need to be increased. The most basic determinant of these thresholds will be whether the system is implemented for convenience or security.

Competing Finger-Scan Technologies

There are a handful of approaches to the problem of acquiring fingerprint images of sufficient quality to create finger-scan templates. Optical, silicon, and ultrasound are the leading methods in use in the finger-scan industry.

Optical Technology

Optical technology is the oldest and most widely used finger-scan technology. The user places a finger on a coated platen, built of hard, coated plastic or coated glass. In essence, a camera registers the image of the fingerprint against a coated glass or plastic platen, upon which the digitized ridges and valleys appear as black, gray, and white lines. The camera acquires a series of images, allowing the underlying software to assess fingerprint quality and generate templates for enrollment or verification.

Optical technology has several strengths: It has been proven reliable over time, is resistant to electrostatic discharge, is fairly inexpensive, and can provide resolutions up to 500 DPI, the benchmark for high-quality fingerprint images. Optical technology's weaknesses include size (the platen must be of sufficient surface area and depth to capture quality images), a sporadic tendency to show latent prints as actual fingerprints, and a susceptibility to fake fingers. Optical devices are deployed in logical access and physical access systems. Most live-scan solutions, devices designed to acquire images for AFIS systems, are based on optical imaging.

Silicon Technology

Silicon technology, which uses a silicon chip as a platen, has gained considerable acceptance since its commercial introduction in 1998. Most silicon finger-scan devices acquire images based on the capacitive characteristics of fingerprint ridges and valleys: The silicon sensor acts as one plate of a capacitor, and the fingerprint becomes the other plate. Silicon technologies can use *active capacitance*, which generates a small field that extends beyond the surface of the platen and reads to the live layer of skin, or *passive capacitance*, which measures up to the point of contact.

Silicon technology's strengths include high image quality, approaching that of the better optical devices; modest size requirements, allowing the technology to be integrated into small, low-power devices; and potentially lower cost, as a large number of silicon platens can be manufactured from a single wafer. On the other hand, silicon's durability is still subject to question. Some types of silicon technology have been susceptible to electrostatic damage, and its performance in challenging conditions is unproven. Silicon devices are deployed almost invariably as logical access solutions, integrated into peripherals for desktop usage. They are infrequently deployed as physical access solutions.

Ultrasound Technology

Ultrasound technology is the least frequently utilized of the three primary finger-scan technologies, but has unique advantages. Ultrasound devices transmit inaudible acoustic waves to the finger, generating images by measuring the impedance between the finger, the platen, and air. Ultrasound devices are more capable of penetrating dirt and residue than optical and silicon devices, and are not subject to some of the image-dissolution problems found in larger optical devices.

A limitation of the technology is the need for a larger acquisition device due to the machinery involved in ultrasonic imaging. As such, today's ultrasonic technology is not well suited for use in desktop applications. Ultrasound technology is most commonly utilized in live-scan devices, acquiring high-quality fingerprint images for automated fingerprint identification systems (AFIS).

Other approaches to finger-scan imaging include thermal imaging, which uses silicon to measure the heat of fingerprint ridges and valleys when swept across a scanner, and pressure-based sensors, which locate ridges and valleys by simply measuring the pressure of a finger placed against a reader.

Feature Extraction Methods: Minutiae versus Pattern Matching

Approximately 80 percent of finger-scan technologies are based on minutiae. A leading alternative is *pattern matching*, which bases its feature extraction and template generation on a series of ridges, as opposed to discrete points. The use of multiple ridges reduces dependence on minutiae points, which tend to be affected by wear and tear, although pattern matching systems seem more sensitive to proper placement of the finger during verification. The templates created in pattern matching are often two to three times larger than minutiae-based templates—approximately 1,000 bytes versus 250 to 500 bytes.

Finger-Scan Deployments

Finger-scan technology is used by hundreds of thousands of people daily to access networks and PCs, to enter restricted areas, and to authorize transactions. In contrast to many biometric technologies that are best suited for use within specific applications, finger-scan can be used successfully in a broad range of applications. The technology is used in a range of vertical markets, including the financial services, healthcare, and government sectors. Notable finger-scan deployments include the following.

The city of Glendale, California, is implementing finger-scan peripherals and accompanying software to all 2,100 city employees, with initial rollout to the Glendale Police Department. The benefits are increased network security, as well as an estimated savings of $50,000 per year on administrative costs due to the increased efficiency and convenience of network access.

Westernbank in Puerto Rico has deployed finger-scan solutions to its 37 branches for customer account access and employee workstation login. The biometric system is available as an option to all customers who wish to use the system. Westernbank representatives stated in mid-2000 that, of approximately 300,000 bank members, 10 to 15 percent opt to use the biometric systems and the rest prefer to use traditional technologies. Westernbank estimates the total cost of the deployment to be approximately $3 million.

In Mexico, Groupo Financiero Banorte, a leading financial institution, has deployed a system of kiosks utilizing smart cards and biometrics for workers who do not hold bank accounts. Using finger-scan technology, the smart card system replaces paychecks, dispensing cash to employees. The smart card also functions as a debit card in local retail establishments. Some 4,000 workers participated in the pilot project, and the full deployment is intended to serve over 650,000 workers.

Since 1998, the New York State Office of Mental Health has purchased and deployed over 6,000 finger-scan units to enhance network security. The systems, used by healthcare staff, are driven in part by HIPAA legislation that demands increased measures to protect the confidentiality of sensitive patient information.

> ## Biometric Middleware
>
> While deployers normally mull their biometric hardware options when considering biometric solutions, in some environments, such as PC/network access, biometric software can be a more difficult and important decision. While many hardware solutions come with dedicated software designed to enable a specific biometric device, a broader solution known as *biometric middleware* is widely available in the marketplace.
>
> Biometric middleware is authentication software that (1) enables various biometric devices and technologies and (2) allows the match or no-match decisions made by core technologies to provide authentication to various PC applications and resources. Middleware solutions may be compatible with as few as 5 and as many as 25 different authentication solutions, with a focus on biometrics but also including smart cards and tokens. Authentication through these devices can provide access to operating systems, applications, or other protected resources. Middleware solutions can be seen as platforms or infrastructures that reduce dependence on a single type of biometric hardware, allowing users to plug new devices into the infrastructure as required.
>
> The basic rationale behind biometric middleware is that enterprises, software developers, and merchants want to integrate biometric functionality into their daily operations, products, or services, but do not want to be tied to a specific biometric device, solution, or technology. Currently, almost all middleware solutions are deployed in employee-facing enterprise implementations. Over time, it is expected that middleware will play a large role in customer-facing applications, as it is nearly certain that home users will have access to a variety of competing hardware solutions—finger-scan, voice-scan, iris-scan, and so on—and merchants will be interested in enabling these solutions. Middleware of some type will be required to bridge the gap between the end user and the merchant.
>
> Microsoft has announced that it plans to incorporate a specific vendor's biometric middleware into its future operating systems, with the probable result that competing middleware offerings will need to focus less on PC/network access and more on customer-facing and transactional applications of their software.

An interesting deployment at Welsh Valley Middle School in Pennsylvania used finger-scan technology to track cafeteria purchases. Instead of paying for food in cash, students have the option of using finger-scan peripherals. At the end of each month, their parents receive a bill for the month's food purchases, or the food bill is sent to a free-food program for reconciliation. The deployment was prompted by legislation that forbids identification of students who participate in federal free-lunch programs. With students paying for lunch via a finger-scan system installed in cafeteria checkout lines, there is no way to determine whether their parents or government grants are paying for their meal.

Finger-Scan Strengths

Finger-scan technology has a number of advantages over competing technologies, some of which are attributable to the core technology, others of which are functions of the finger-scan marketplace.

Proven Technology Capable of High Levels of Accuracy

The fingerprint has long been recognized as a highly distinctive identifier; classification, analysis, and study of fingerprints have existed for decades. The combination of an innately distinctive feature with a long history of use as identification sets finger-scan apart in the biometric industry. There are physiological characteristics more distinctive than the fingerprint (the iris and retina, for example), but automated identification technology capable of leveraging these characteristics has been developed only over the past few years, not decades.

Strong finger-scan solutions are capable of processing thousands of users without allowing a false match and can verify nearly 100 percent of users with one or two placements of a finger. In addition, the technology can be deployed in a one-to-few fashion in which individuals are matched against modest databases, typically of 10 to 100 users. Many finger-scan solutions can be deployed in applications where both security and convenience are primary drivers.

Range of Deployment Environments

Reduced size and power requirements, along with finger-scan's resistance to environmental changes such as background lighting and temperature, allow the technology to be deployed in a range of logical and physical access environments. Finger-scan acquisition devices have grown quite small—devices slightly thicker than a coin and less than 1.5 cm x 1.5 cm are capable of acquiring and processing images.

In addition, there are more finger-scan solutions in the biometric marketplace than all other technologies combined. Though this may be seen as a weakness—the would-be deployer or customer may be overwhelmed by the number of solutions available—competition in the marketplace has resulted in a number of robust solutions for desktop, laptop, physical access, and point-of-sale environments.

Ergonomic, Easy-to-Use Devices

The act of placing a finger on a device is largely intuitive and can be grasped with little training. Many other biometric technologies require complex user-system interactions, while finger-scan devices are generally designed such that placement is an easily repeatable process. In addition, the design of finger-scan devices has improved substantially. The first finger-scan devices were bulky and did not lend themselves to consistent finger placement, whereas many of today's devices guide the finger to proper placement.

Ability to Enroll Multiple Fingers

Although most deployers do not capitalize on this functionality, the fact that most people can enroll up to 10 fingers in a biometric system gives finger-scan advantages in security and flexibility. For example, requiring placement of two fingers in succession can make for a much more secure biometric system—the odds against false matching against two separate fingers are astronomical. Allowing a user to verify with one of several enrolled fingers reduces the likelihood of a user being falsely rejected. Users might enroll certain fingers with specific functionality—an alarm finger for a physical access deployment if under duress, or a specific finger used to log into specific Web sites or applications. To reduce the likelihood of biometric information being stolen and reused, a system may challenge the user to put down one of several fingers—only the correct response would be accepted.

Finger-Scan Weaknesses

The following weaknesses affect nearly all finger-scan solutions, but can be mitigated through intelligent system design.

Inability to Enroll Some Users

A small percentage of users—anywhere from a fraction of a percent to as high as several percent, depending on the technology and user population—are unable to enroll in many finger-scan systems. Certain ethnic and demographic groups have lower-quality fingerprints and are more difficult to enroll than others. IBG's Comparative Biometric Testing has shown that elderly populations, manual laborers, and some Asian populations are more likely to be unable to enroll in some finger-scan systems.

In an enterprise deployment for physical or logical security, this means that some number of users need to be processed by another method, be it another biometric, a password, or a token. In a customer-facing application, this may mean that a customer willing to enroll in a new system is simply unable to. In a large-scale 1:N application, the result may be that a user is able to enroll multiple times because his or her fingerprints cannot be reliably identified. If the system is designed to be more forgiving and to enroll even questionable fingerprints, then the common result is increased error rates.

Performance Deterioration over Time

Although the fingerprint is a stable physiological characteristic, daily wear can cause the performance of some finger-scan technologies to drop drastically. Some systems' error rates have gone from nonexistent to 25 percent over a span of six weeks, while others remain unchanged. Although high-quality enrollment improves long-term performance, users who work with their hands are likely to see increased error rates over time.

Institutions considering implementation of finger-scan solutions must ensure that the solutions deployed are capable of verifying users over time and that enrollment procedures reduce the likelihood that performance will deteriorate. Enrolling multiple fingerprints and reenrolling users whose ability to verify decreases over time are two ways to overcome this weakness.

Association with Forensic Applications

Finger-scan technology's similarity to forensic fingerprinting causes some percentage of users discomfort. Although finger-scan systems do not store images and are, by and large, incapable of the type of large-scale searches used in systems, the stigma attached to fingerprints may eventually limit the growth and acceptance of finger-scan technology. Privacy advocates fear that finger-scan data collected for a specific purpose may be used for forensic applications or to track a person's various activities. Regardless of the viability of these fears, the general public's association of finger-scan technology with criminal fingerprinting cannot be ignored.

Need to Deploy Specialized Devices

In order for finger-scan to become a pervasive biometric solution, devices must be present on desktops, at points of sale, at protected doorways, and at any location where authentication is required. As opposed to voice-scan, for example, which can leverage a widely available infrastructure of telephones and microphones, finger-scan requires that specialized acquisition devices be installed. The integration of finger-scan devices within keyboards, for example, would begin to address this challenge for desktop deployments, but would not address the challenge of deploying devices for use in physical access.

Finger-Scan: Conclusion

Finger-scan technology is a strong solution for a range of environments and will likely continue to grow in prominence in logical and physical access applications. It is not an ideal solution for every situation, and technologies with fewer negative associations may be more acceptable to some potential users. However, finger-scan will likely be central to the biometric industry's growth, as it is capable of being effectively deployed in the widest range of authentication environments.

Deployers will face the challenge of deciding between dozens of finger-scan systems, some of which operate only with their own technology, others of which are compatible with various middleware vendors. As standards are adopted more widely in the industry, there will be more flexibility in initial device selection and implementation.

For a description of the latest finger-scan vendors and technologies, visit www.biometricgroup.com/wiley.

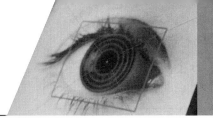

CHAPTER 5

Facial-Scan

Facial-scan technology utilizes distinctive features of the human face in order to verify or identify individuals. Facial-scan currently plays a role in the biometric marketplace in 1:N identification applications. Used in conjunction with ID card systems, in booking stations, and for various types of surveillance operations, facial-scan's most successful implementations take place in environments where cameras and imaging systems are already present. Facial-scan is also deployed in select environments as a 1:1 verification solution for physical and logical access, but has found only limited implementation in these areas.

Facial-scan's strengths include the following:

➢ It has the ability to leverage existing image acquisition equipment.
➢ It can search against static images such as driver's license photographs.
➢ It is the only biometric able to operate without user cooperation.

Facial-scan's weaknesses include the following:

➢ Changes in acquisition environment reduce matching accuracy.
➢ Changes in physiological characteristics reduce matching accuracy.
➢ It has the potential for privacy abuse due to noncooperative enrollment and identification capabilities.

Components

Facial-scan systems can range from software-only solutions that process images acquired through existing closed-circuit television (CCTV) cameras to full-fledged acquisition and processing systems, including cameras, workstations, and back-end processors. In some facial-scan systems, the core technology is optimized to work with specific cameras and acquisition devices. More often, the core technology is designed to enroll, verify, and identify facial images acquired through various methods—static photographs, Web cameras, surveillance cameras. The two basic components of facial-scan technology—the face location engine, which finds and tracks faces in a field of view, and the face recognition engine, which compares these faces—are normally located on the same PC or device, working in close conjunction. Facial-scan template matching normally takes place on a local or central PC, although the core technology has been embedded within devices such as personal digital assistants (PDAs) and mobile phones (see Figure 5.1).

Facial-scan systems are not often integrated into 1:1 logical or physical access applications and are more likely to be used in large-scale identification or surveillance. The need to perform complicated postmatch integration

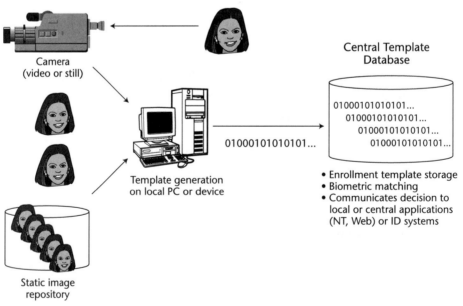

Figure 5.1 Components of a facial-scan system.

functionality is therefore less present in facial-scan systems. As opposed to a finger-scan system, where a match is normally integrated directly into some existing system, a facial-scan match often triggers a manual process of investigation. The results of facial-scan matches are less likely to tie into automated processes or functions than other biometrics.

How Facial-Scan Technology Works

Facial-scan technology is based on the standard biometric sequence of image acquisition, image processing, distinctive characteristic location, template creation, and matching.

Image Acquisition

Facial-scan technology can acquire faces from almost any static camera or video system that generates images of sufficient quality and resolution. Ideally, images acquired for facial-scan will be acquired through high-resolution cameras, with users directly facing the camera and with moderate lighting of the face. High-quality enrollment is essential to eventual verification and identification; enrollment images define the facial characteristics to be used in all future authentication events. Although certain technologies can search digitized images for faces as small as 30 pixels in height, the technology performs much better when the height and width of the facial images each exceeds 100 pixels.

Image acquisition becomes more difficult and the technology's accuracy becomes less robust under many circumstances. Distance from camera reduces facial size and therefore image resolution, though the technology is capable of magnifying smaller faces to increase the possibility of template generation. Users not looking directly at the camera—positioned more than approximately 15 degrees—either vertically or horizontally, away from ideal positioning—are less likely to have images acquired by some technologies. There is also a gap between the threshold for locating faces in a field of view and locating usable faces—a system may be able to locate a face from a sharp angle, but be unable to subsequently enroll or identify the individual from this image. To overcome the acquisition-angle problem, newer facial-scan methods actually require that the user look left, right, up, and down in order to acquire images from various angles, thus building a more comprehensive enrollment template.

While angled acquisition is being addressed in certain facial-scan technologies, the problem of lighting is more severe. While the human brain is capable of acquiring facial images under drastically different lighting conditions, facial-scan technologies are generally unable to acquire images that are somewhat overexposed or underexposed. Systems with automatic gain control, able to adjust for different lighting conditions and skin tone, are more likely to present images from which facial-scan solutions can acquire faces. Facial-scan systems' sensitivity to lighting and gain can actually result in reduced ability to acquire faces from individuals of certain races and ethnicities. Select Hispanic, black, and Asian individuals can be more difficult to enroll and verify in some facial-scan systems because acquisition devices are not always optimized to acquire darker faces. At times, an individual may stand in front of a facial-scan system and simply not be found. While the issue of failure-to-enroll is present in all biometric systems, many are surprised that facial-scan systems occasionally encounter faces they cannot enroll.

The severity of these image acquisition challenges applies in varying degrees to facial-scan systems implemented for 1:1 verification, for 1:N public-sector identification, and for surveillance. Generally speaking, 1:N public-sector identification systems are most likely to have controlled and consistent enrollment environments: Users can be required to stand at a fixed distance from a camera, with fixed lighting and a fixed background. In 1:1 verification, variables such as changes in lighting and acquisition angle are likely to be introduced, and the lack of supervision in most 1:1 verification implementations exacerbates the problem. Surveillance is a worst-case scenario as images are acquired from unaware subjects, moving at various speeds and angles, through cameras that generally lack active gain control. Because initial image acquisition is critical to system operations, deployers must be sure to take every variable into account when determining how and where to implement facial-scan technology (see Figure 5.2).

Image Processing

Once the primary challenge of facial-scan systems—image acquisition—is addressed, image processing follows. Images are cropped such that the ovoid facial image remains, and color images are normally converted to black and white in order to facilitate initial comparisons based on grayscale characteristics. Facial images are then normalized to overcome variations in orientation and distance. In order to do this, basic characteristics such as the middle of the eyes are located and used as a frame of reference. Once the eyes are located, the facial image can be rotated clockwise or counterclockwise to straighten the image along a horizontal axis. The face can then be magnified, if necessary, so that the facial image occupies a minimum pixel space. Since most facial-scan systems acquire multiple face images to enroll individuals—from as few as 3 images to well over 100, depending on the vendor and the matching method—rapid image processing routines are essential to system operations.

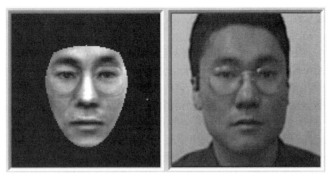

High-Quality Enrollment Image

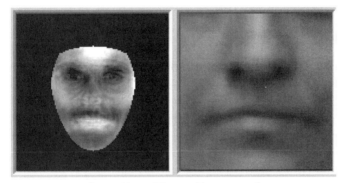

Low Quality Enrollment Image

Figure 5.2 High-quality and low-quality facial-scan images.

Distinctive Characteristics

Once an image is standardized according to the vendor's requirements, the core processes of distinctive characteristic location can occur. While there are various matching methods, all facial-scan systems attempt to match visible facial features in a fashion similar to the way people recognize one another. The features most often utilized in facial-scan systems are those least likely to change significantly over time: upper ridges of the eye sockets, areas around the cheekbones, sides of the mouth, nose shape, and the position of major features relative to each other. Areas that are very likely to change or to be obscured, such as areas of the face immediately adjacent to a hairline, are usually not relied upon for verification. See the section entitled *Competing Facial-Scan Technologies* for discussion of the leading facial-scan methodologies.

One of the challenges involved in facial-scan technology is that the face is a reasonably changeable physiological characteristic. As opposed to a fingerprint, which might be scarred but is difficult to alter dramatically, faces can be changed enough to reduce a system's matching accuracy. A user who smiles during enrollment and grimaces during verification or identification is more likely to be rejected than one who does not intentionally alter his or her expression during authentication. Behavioral changes such as alteration of hairstyle, changes in makeup, growing or shaving facial hair, adding or removing eyeglasses are behaviors that impact the ability of facial-scan systems to locate distinctive features. While the human brain may be able to account for many of these changes, facial-scan systems are not yet developed to the point where they can overcome such variables.

Template Creation

Enrollment templates are normally created from a multiplicity of processed facial images. These templates can vary in size from less than 100 bytes, generated through certain vendors' quick-search algorithms, to over 3K for templates created through full search algorithms. These templates cannot be used to recreate original images, but are instead proprietary representations of the data located during feature location. The 3K template is by far the largest among technologies considered physiological biometrics. Larger templates are normally associated with behavioral biometrics, wherein it is difficult to locate distinctive characteristics with precision.

Template Matching

Vendors employ proprietary methods to compare match templates against enrollment templates, assigning confidence levels to the strength of each match attempt. If the score surpasses a predefined level, the comparison is deemed a match. In many cases, a series of images is acquired and scored against the enrollment, so that a user attempting 1:1 verification within a facial-scan system may have 10 to 20 match attempts take place within 1 to 2 seconds. This sets facial-scan apart from most other biometrics, in which a single match template is acquired and scored, and, if necessary, the user is prompted to attempt again. Whereas in most 1:1 biometric systems a rejection is defined as a failure to match after a given number of attempts (often 1 to 3), a rejection in a facial-scan system is often defined as a failure to match after a certain amount of time.

In identification, the number of match attempts and the rate of matching will increase dramatically. Leading facial-scan systems can perform tens of thousands of identification comparisons per second through standard PCs. These rapid searches often take place through the smaller template generated in facial-scan systems. Because facial-scan is not as effective as finger-scan or iris-scan in identifying a single individual from a large database, a number of potential matches are generally returned after large-scale facial-scan identification searches. For example, a system may be configured to return the 10 most likely matches on a search of a 10,000-person database. A human operator would then determine which if any of the 10 potential matches is an actual match.

Competing Facial-Scan Technologies

While the internal operations of a facial-scan system are invisible to the deployer, whose primary concern is performance and accuracy, a handful of facial-scan technologies compete within the biometric market, with substantial differences in their operations. Because of their enrollment or verification methods, some types of facial-scan technology are more suitable than others for applications such as forensics, network access, and surveillance. Four of the primary methods employed by facial-scan vendors to identify and verify subjects include Eigenface, feature analysis, neural network, and automatic face processing. Other facial-scan technologies based on thermal patterns present under the skin have not yet proven commercially viable.

Eigenface

Eigenface, roughly translated as "one's own face," is a technology patented at MIT that utilizes a database of two-dimensional, grayscale facial images (Eigenfaces) from which templates are created during enrollment and verification. These Eigenfaces feature distinctive facial characteristics, and the vast majority of faces can be reconstructed by locating distinctive features from approximately 100 to 125 Eigenfaces. Variations of Eigenface are frequently used as the basis of other face-recognition methods (see Figure 5.3).

Upon enrollment, a subject's facial image is represented using a combination of various Eigenfaces. This reconstruction is then mapped to a series of numbers or coefficients. For 1:1 authentication, in which the image is being used to verify a claimed identity, an individual's live template is compared against the

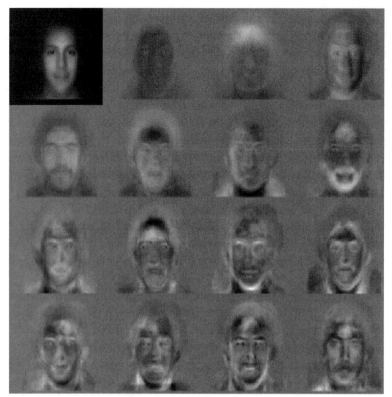

Figure 5.3 Eigenfaces.

enrolled template to determine coefficient variation. The degree of variance from the enrollment will determine acceptance or rejection. For 1-to-many identification, the same principle applies, but with a much larger comparison set. Like all facial recognition technology, Eigenface technology is best utilized in well-lit, frontal image capture situations.

Feature Analysis

Feature analysis is perhaps the most widely utilized facial recognition technology. This technology is related to Eigenface, but is more capable of accommodating changes in appearance or facial aspect (smiling versus frowning, for example). Visionics, a prominent facial recognition company, uses Local Feature Analysis (LFA), which can be summarized as a reduction of facial features to an "irreducible set of building elements."

Feature analysis derives enrollment and verification templates from dozens of features from different regions of the face and also incorporates the relative location of these features. The extracted features are building blocks, and both the type of blocks and their arrangement are used for identification and verification. It anticipates that the slight movement of a feature located near one's mouth will be accompanied by relatively similar movement of adjacent features. Since feature analysis is not a global representation of the face, it can accommodate angles up to approximately 25 degrees in the horizontal plane, and approximately 15 degrees in the vertical plane. A straight-ahead facial image from a distance of 3 feet will be the most accurate.

Neural Network

Neural network systems employ algorithms to determine the similarity of the unique global features of live versus enrolled or reference faces, using as much of the facial image as possible. Neural systems are designed to learn which features are most effective within the body of users that the system is intended to serve. Features from both the enrollment and the verification faces vote on whether there is a match. An incorrect vote, such as a false match, prompts the matching algorithm to modify the weight it gives to certain facial features. In this way, neural network systems learn which features are most effective for matching and pragmatically adjust themselves based on the methods that have proven most effective. This method, theoretically, leads to an increased ability to identify faces in difficult conditions.

Other facial technologies have emerged based on more advanced neural models, with detailed cells incorporating thousands of facial images. Since these technologies are capable of learning over time, they may be capable of reducing the time-based performance problems found in many facial-scan systems. However, their elongated enrollment process means that they are not well-suited for surveillance applications in which users are matched against watch lists. These watch lists are often generated from static images, not the ideal environment for neural net enrollment.

Automatic Face Processing

Automatic face processing (AFP) is a more rudimentary technology, using distances and distance ratios between easily acquired features such as eyes, end of nose, and corners of mouth. Though overall not as robust as Eigenfaces, feature analysis, or neural network, AFP may be more effective in dimly lit, frontal image-capture situations. It is often used in booking station applications in which environmental conditions are more controlled.

Facial-Scan Deployments

Facial-scan is generally deployed in environments where existing acquisition technology or facial images are in place, such as public-sector ID card applications, surveillance systems, and booking stations. In ID card scenarios, facial-scan leverages existing photographs and has been used for fraud detection and deterrence in voter registration, benefit disbursement, and driver's license applications. These situations are ideal for facial-scan because both enrollment and verification image acquisition take place in similar, controlled environments. Notable facial-scan deployments include the following.

The first public deployment of surveillance technology in Newham, England, uses 144 CCTV cameras to record all activity on several streets in what had been an unsafe neighborhood. The faces acquired through the cameras are compared to a hotlist of known offenders using Visionics facial-scan technology. Largely a deterrent to street crime, the system has reduced the incidence of various offenses by 21 to 39 percent.

The Mexican government has licensed Visionics facial-scan technology for voter registration. After its one-time usage in the July 2000 Mexican national elections, a permanent license for the technology was acquired. The purpose of the system is to deter and prevent citizens from voting more than once under different aliases.

Biometrica operates independently as a specialized integrator of casino surveillance systems using Viisage technology. It has installed systems in over 70 casinos worldwide, including Foxwoods Casino (Connecticut), the Trump Marina, Taj Mahal, and Plaza (Atlantic City, New Jersey), the Stratosphere Hotel and Casino and the Mirage Resort (Las Vegas, Nevada). Similarly, Griffin Investigations has deployed Visionics technology in numerous casinos worldwide. Griffin Investigations maintains a database of casino offenders for use by participating casinos and law enforcement entities, and has partnered with Visionics to create an application that can leverage and increase the functionality of this database.

Illinois contracted Viisage to provide a driver's license system incorporating facial-scan technology for its 25 million residents. The system performs 1:N matching against exiting users to ensure that duplicate licenses are not issued.

Facial-Scan Strengths

Facial-scan has a handful of key strengths that distinguishes it from most other biometrics.

Ability to Leverage Existing Equipment and Imaging Processes

Unlike many other physiological biometrics, facial-scan is a software-based technology that can be deployed without the addition of proprietary hardware or extensive retrofitting. In many cases, the technology is capable of using existing imaging systems such as standard video cameras. The large number of surveillance and CCTV cameras in use, as well as the wide variety of photo ID systems deployed at state and national levels, provides a built-in infrastructure that facial-scan technology can leverage. Performance does increase with higher resolutions and frame rates, as better facial images are acquired.

From a deployer's perspective, the ability to deploy facial-scan without the need to introduce a range of new policies or equipment allows comparatively low-cost and low-effort project initiation.

Ability to Operate without Physical Contact or User Complicity

Facial-scan is the only biometric capable of identification at a distance without subject complicity or awareness. This allows facial-scan to operate in surveillance mode; many casinos have adopted facial-scan systems into their already existing CCTV systems in an effort to identify cardsharps. Police and government agencies have also installed facial-scan systems in various public places to prevent and deter crime.

Some users express concerns about using biometric devices that involve physical contact, such as finger-scan or hand-scan. There may be discomfort about touching a device that others have touched—users occasionally ask about germs, and so on. Facial-scan is not subject to this type of opposition.

Ability to Enroll Static Images

One of the major challenges of a large-scale biometric system is initial enrollment. Because of the logistics involved in projects involving more than 1 million users, it can take years to enroll the target population. Many civil Automated Fingerprint Identification Systems (AFIS) projects, for example, are deployed over a 3-to-5-year timeframe. In large-scale facial-scan deployments, there is often an existing database of usable facial images collected in controlled environments over time.

The same applies to surveillance applications. If an individual has had one or more high-quality photographs taken of him- or herself, this can in many cases serve as an enrollment in a facial-scan system. These users would be added to a watch list against which surveillance searches take place. There are effectively tens of millions of latent facial-scan images in departments of motor vehicles (DMVs), immigration offices, and other public agencies around the world.

Facial-Scan Weaknesses

There are a handful of limiting factors that reduce facial-scan's effectiveness in a range of environments.

Acquisition Environment Effect on Matching Accuracy

The accuracy of facial-scan solutions drops significantly under certain enrollment and verification conditions. In order to enroll successfully, users must be facing the acquisition camera and cannot be acquired from sharp horizontal or vertical angles. The user's face must be lit evenly, preferably from the front. Though these are not problems for applications such as ID cards, in which lighting, angle, and distance from camera can be controlled, applications where environmental factors vary are much more difficult for facial-scan systems.

Facial-scan systems are especially ineffective when users are enrolled in one location and verified in another. Factors such as direct and ambient lighting, camera position and quality, angle of acquisition, and background composition can dramatically reduce accuracy. Users must enroll and verify under the same conditions for the technology to operate effectively. Reduced accuracy is most strongly reflected in false nonmatches—authorized users being incorrectly rejected by the facial-scan system.

Changes in Physiological Characteristics That Reduce Matching Accuracy

In addition to the impact that environmental changes have on accuracy, simple changes in user appearance seem to have an impact on many systems' ability to verify users. Changes in hairstyle, makeup, or facial hair, addition or removal of eyeglasses, even hats and scarves can cause users to be falsely rejected. Discussions of whether facial-scan systems can distinguish between identical twins or whether plastic surgery is necessary to evade detection seem out of place when such basic changes can impact the system's operations.

Over time, it is very likely that improvement in the technology will increase its ability to match users under more challenging conditions. The face is clearly a distinctive characteristic and indeed comprises several distinctive elements (nose, eye shape, lips, etc.). Advances in the core technology that enables acquisition at sharper angles and can model and verify faces in three dimensions should help reduce the technology's propensity toward false nonmatching.

Potential for Privacy Abuse Due to Noncooperative Enrollment and Identification

While there are benefits to facial-scan's ability to operate covertly, the potential for privacy-invasive misuse is evident. A facial-scan system capable of 1:N operation, able to acquire images without consent, could become the tracking system feared by many opponents of biometrics. Certain facial-scan deployments have met with public objections—in particular, two controversial deployments in Florida: one aimed to prevent crime in a popular shopping district and one aimed to catch criminals at the 2001 Super Bowl. Similarly, a facial-scan system in Uganda, established to prevent voter fraud, was heavily criticized as intimidating voters opposed to the existing regime. Potential future utilization of facial-scan could be severely limited if people come to view the technology as a tool that invades privacy.

Facial-Scan: Conclusion

Facial-scan has unique advantages over all other biometrics in terms of surveillance operation, ability to leverage millions of preexisting images, and ability to leverage existing acquisition devices. However, the core technology is highly susceptible to falsely nonmatching users in 1:1 verification mode and to failing to identify enrolled users in 1:N mode. Factors such as environmental changes and mild changes in appearance impact the technology to a greater degree than many expect. For environments in which the very presence of a biometric system is adequate to effect some goal, such as deterrence, facial-scan technology can be a highly useful technology. For implementations where the biometric system must verify and identify users reliably over time, facial-scan can be a very difficult—but not impossible—technology to implement successfully.

For a description of the latest facial-scan vendors and technologies visit www.biometricgroup.com/wiley.

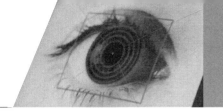

CHAPTER 6

Iris-Scan

Iris-scan technology utilizes the distinctive features of the human iris in order to identify or verify the identity of individuals. Iris-scan technology has the potential to play a large role in the biometric marketplace if real-world systems and solutions meet the technology's theoretical promise. Traditionally having been used in high-security physical access applications, iris-scan technology has been successfully implemented in ATMs and kiosks for banking and travel applications and is being positioned for desktop usage.

Iris-scan's strengths include the following:

- It has the potential for exceptionally high levels of accuracy.
- It is capable of reliable identification as well as verification.
- It maintains stability of characteristic over a lifetime.

Iris-scan's weaknesses include the following:

- Acquisition of the image requires moderate training and attentiveness.
- It has a propensity for false rejection.
- A proprietary acquisition device is necessary for deployment.
- There is some user discomfort with eye-based technology.

Components

Iris-scan systems comprise front-end acquisition hardware along with local or central processing software. As opposed to facial-scan systems, which can leverage existing camera technology, iris-scan deployments require specialized devices that provide necessary infrared illumination. These devices can take the form of desktop cameras (with video camera functionality along with iris-scan functionality) or more robust cameras designed for integration into physical access units and kiosks (see Figure 6.1).

The software components of an iris-scan system—the image processing and matching engines and the proprietary database—can reside on a local PC attached to the acquisition unit or be spread between a local and a central PC. In implementations where a series of iris-scan peripherals are deployed, such as in a network security implementation, a central server is generally used to match and store data, while the local PCs will perform template generation. In this case, what is transmitted across internal or external networks is a template, not an identifiable image. Newer iris-scan solutions are Web-enabled, such that a widely dispersed implementation could share a single central database.

The results of iris-scan matches are generally tied directly into logical or physical access systems, resulting in access to protected resources being granted or

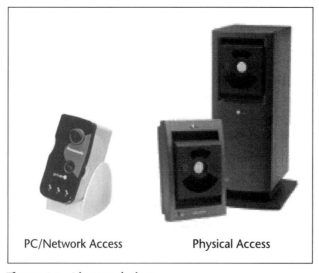

Figure 6.1 Iris-scan devices.

restricted. The main provider of iris-scan technology works with various middleware vendors and hardware integrators who provide this integration logic to existing systems, replacing existing authentication methods.

How It Works

Because only one company provides the underlying biometric technology, there are fewer variables involved in the iris-scan's image and template-oriented functions than in other biometrics.

Image Acquisition

Iris-scan technology requires the acquisition of a high-resolution image of the eye, illuminated by an infrared imager, in order to effectively map the details of the iris. The acquisition process and the amount of effort required on the part of the user differ according to the type of acquisition device used. The three major types of iris-scan systems are kiosk-based systems, physical access devices using motorized cameras, and inexpensive desktop cameras. Although iris-scan vendors do not emphasize their use of infrared light, each system does rely on infrared imaging using wavelengths in the 700- to 900-nm range (judged to be safe by the American Academy of Ophthalmology).

Kiosk systems require that users stand approximately 2 to 3 feet from the camera or cameras, which are positioned at the height of a typical user's eyes. The user must remain still, as it is difficult for these systems to locate irises on a moving target. When faced with a user in its field of view, the kiosk-based camera searches for eye shapes. In order to facilitate this, users may need to remove their eyeglasses, because glare from eyeglasses can impact the ability of systems to match users. From this point, the acquisition is automatic—the system normally locates the iris and acquires the image within 1 to 2 seconds.

Physical access devices require slightly more user effort. A small camera mounted behind a mirror acquires the image; the user locates his or her eye in the mirror, centering the iris within a 1-inch by 1-inch square. The user may also be vocally prompted to move slightly forward or backward to enable image capture. The proper distance from the mirror is approximately 3 inches. A high-quality camera focuses on the eye, acquiring a series of images until a resolution threshold is met. Acquisition through these devices is more challenging because it is contingent on users' ability to follow interactive prompts. Also, individuals who favor a particular eye can have problems locating their weaker eye in the mirror.

Desktop cameras, used for logical access, are the newest type of iris-scan device. Acquiring the image at a distance of approximately 18 inches, the device requires the user to align his or her line of sight with a guidance light or hologram. When the user is positioned correctly, the camera acquires the image. These systems have proven fairly difficult for some users, who find it difficult to orient themselves at the proper distance from the camera.

Image Processing

Regardless of the acquisition device, the process of mapping the iris remains the same. After the camera locates the eye, an algorithm narrows in from the right and left of the eye to find the iris's outer edge. The iris-scan algorithm then locates the inner edge of the iris at the pupil. Locating the iris-pupil border can be challenging for users with very dark eyes, as there may be very little difference in color as rendered in the technology's 8-bit grayscale imaging.

Once the parameters of the iris have been defined, a black-and-white image of the iris is used for feature extraction. The core technology can account for pupil dilation, occlusion due to eyelids, and reflections due to the acquisition camera; the area used for feature extraction is a horizontal band extending from the far left to the far right of the iris. When the pupil dilates, the iris patterns shrink and expand in a normalized fashion such that algorithms can translate a dilated verification to a nondilated enrollment.

Distinctive Features

The patterns that constitute the visual component of the iris are surprisingly distinctive. A primary visible characteristic is known as the *trabecular meshwork*, a tissue that gives the appearance of dividing the iris in a radial fashion. Other visible characteristics include rings, furrows, freckles, and the corona. Iris patterns are formed before birth and remain stable throughout an individual's lifetime (unless one suffers an eye injury). Tests have shown that individuals' left and right eyes have different iris patterns, and that even identical twins' irises have almost no statistical similarity.

Iris-scan algorithms map segments of the iris into hundreds of independent vectors. The characteristics derived from iris features are the orientation and spatial frequency of distinctive areas (the *what*) along with the position of these areas (the *where*). Not all of the iris is used; a portion of the top as well as 45 degrees of the bottom are unused to account for eyelid occlusion and reflections (see Figure 6.2).

Figure 6.2 Iris images.

Template Generation

The vectors located by the iris-scan algorithm are used to form enrollment and match templates, which are generated in hexadecimal format as opposed to binary. Depending on the iris-scan solution, between one and four iris images may need to be captured for enrollment template generation. The use of multiple images ensures that the data extracted to form a template is consistent and that there are no reflections being misinterpreted as iris features (see Figure 6.3).

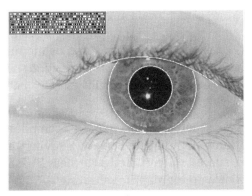

Figure 6.3 An IrisCode.

Template Matching

Iris-scan solutions generally perform identification as opposed to verification, meaning that the match template is compared against all system enrollments to find the best match. Although identification is much more challenging than verification, the process is normally very brief: On some processors, iris-scan technology is capable of searching hundreds of thousands of records per second. Because of the distinctive nature of the iris, identification deployments are not designed to return candidate lists but high-confidence matches.

Though some deployments by their nature must perform identification as opposed to verification, there are risks involved in implementing iris-scan systems in identification mode when verification is possible. In a verification system, a user must claim an identity before interacting with the system. If falsely accepted, he or she has already made the decision to claim another's identity, and the user can easily be held culpable for ensuing system misuse. In identification deployments, there is no identity claim. A falsely matched user may be granted access to resources to which he or she is not entitled, but the lack of an identity claim reduces the person's culpability. While highly resistant to false matching, iris-scan technology is not immune to this happening.

Deployments

While iris-scan technology has had many successful implementations in physical access and kiosk-based identification, it has not been deployed widely in network security or PC-oriented applications. Developers of the iris-scan technology expect that network security will become a major growth area for iris-scan.

Prisons in Pennsylvania, New York, and Florida use iris-scan technology to verify the identity of convicts before release. This system is designed to avoid the possibility of accidentally releasing the wrong prisoner as the result of a fraudulent identity claim. Other implementations in correctional facilities enroll visitors to ensure that people leaving the facilities are visitors, not inmates.

Iris-scan was one of the first biometric technologies to have been piloted in an ATM environment. In 1997, Nationwide Building Society, a savings and loan institution in Swindon, England, successfully piloted a cardless iris-scan ATM using Sensar technology. The pilot won high approval ratings—94 percent of customers preferred iris-scan to PINs. Similar customer-facing pilots took place at Bank United in Texas and Takefuji Bank in Japan. In Frankfurt, Germany, Dresdner Bank has been conducting an iris-scan ATM pilot since late

1999. Thus far, only bank employees have been given the opportunity to use the system, with about 1,000 enrolled as of late 2000.

In July 2000, using technology licensed from Iridian (then IriScan), EyeTicket integrated an iris-scan system in North Carolina's Charlotte/Douglas International Airport. Iris-scan devices are used to control the movement of employees into secure areas, replacing traditional access technologies. A similar solution was deployed in Frankfurt, Germany, in early 2001. Both projects have been described as highly successful, with no false matches having occurred.

A more ambitious implementation involving passenger processing was initiated in July 2001 at London's Heathrow Airport. This implementation allows selected frequent travelers on British Airways and Virgin Airways to clear immigration through iris-scan verification. This is the first use of iris-scan technology in passenger-facing implementation. A similar project was announced in Canada, under the Canadian Airports Council Expedited Passenger Processing System Project. Frequent passengers at eight major Canadian airports may elect to use travel cards storing a biometric template. Cards will allow the passengers to verify their identity at unattended kiosks to simplify border crossings.

Iris-Scan Strengths

Iris-scan's strengths are well-known among those interested in biometrics, although many of these strengths are based on theoretical capabilities as opposed to being derived from real-world experience.

Resistance to False Matching

There is a tremendous amount of distinctive data in an iris, and this data differs substantially from user to user, even between a user's left and right eyes. The algorithms that convert these characteristics into templates generate the most distinctive templates in the biometric industry when given an iris image of sufficiently high resolution.

Although many more iris-scan deployments in logical and physical access must be evaluated before coming to any full conclusions about real-world accuracy, testing and early-phase deployments have shown iris-scan to be extremely resistant to false matching. Even highly accurate technologies such as finger-scan, for which the likelihood of a false match in a 1:1 environment may be 1/100,000, do not approach the false match performance of iris-scan technology. Resistance to false matches positions iris-scan as a strong technology for high-value transactions in both physical and logical access.

There have been at least two incidents, one on controlled testing, one in an operational environment, in which users were incorrectly identified. For them to be most meaningful, false match rates must be balanced with a system's false nonmatch rates and failure-to-enroll rates. Iris-scan technology will need to prove capable of addressing needs in all three areas.

Stability of Characteristic over Lifetime

Although other technologies are capable of highly accurate identification, the iris is unique in that it does not change over a person's lifetime. Other technologies may require reenrollment after a period of time to ensure effective operation, as the physiological characteristic may have changed over time. This resistance to environmental changes is more important than many deployers realize, as reenrollment poses logistical problems and normally follows a long period of deteriorating performance.

The distinctiveness of the iris, when acquired at high resolution, could make iris-scan suitable for use in large-scale civil ID applications in which both identification and verification are necessary. Though a variety of barriers would need to fall before this becomes possible, iris-scan does have intractable advantages over technologies such as finger-scan and facial-scan currently used in large-scale ID applications.

Suitability for Logical and Physical Access

Iris-scan technology was, for most of its history, used for physical access and the occasional ATM solution. Advances in capture technology have reduced the size of iris capture devices to that of a standard video camera. Whereas most technologies are clearly better suited for logical or physical access, iris-scan technology now has the potential to reach both markets effectively. The fact that both physical and logical access solutions are based on the same vendor's core technology may mean that systems can be developed to offer integrated desktop and building access.

Iris-Scan Weaknesses

Iris-scan weaknesses pertain mainly to operational issues: the difficulty of usage, the propensity for false rejections, user perception of eye-based technologies, and the need for a specialized acquisition device.

Difficulty of Usage

Iris-scan, though not the most difficult biometric to use, is not yet as intuitive as technologies such as finger-scan and facial-scan. Users must be cognizant of the manner in which they interact with the system, as enrollment and verification require fairly precise positioning of the head and eyes. Also, users with poor eyesight and those incapable of lining up their eye with the technology's guidance components have difficulty using the technology.

Iris-scan's ease of use varies by device. Desktop devices have proven to be among the hardest to use, as individuals must position themselves without knowing precisely how far away from the camera they should be. Physical access devices are slightly easier because the distance to the acquisition device is shorter. Kiosk-based solutions, which employ the most advanced camera technology, are the easiest to use but still require compliant and attentive users.

False Nonmatching and Failure-to-Enroll

Directly related to the difficulty of acquiring acceptable iris images are the problems of false rejection, or false nonmatching, and failure-to-enroll. Testing has shown that a fair percentage of users are unable to enroll in iris-scan systems because they cannot provide adequate enrollment images. Testing has also shown that, on a fairly frequent basis, enrolled users cannot be identified from databases of modest size. These two performance problems are difficult to solve simultaneously; in order to reduce the failure-to-enroll rate, one can implement systems with lower enrollment thresholds, allowing marginal images to be used for enrollment. This leads to increased false nonmatch rates, as users have difficulty being identified against marginal enrollments.

While the capabilities of iris-scan technology processing images in a laboratory environment have been shown to be quite impressive, this has not always translated into real-world effectiveness.

User Discomfort with Eye-Based Technology

Although there may be little rational justification for their response, a substantial percentage of users are uncomfortable with the idea of using an eye-based biometric. For some, it seems simply to be a visceral reaction, a squeamishness about the eye. For others, there is the concern that exposure to the technology may damage eyesight. These reactions are present even when users are unaware that iris-scan technology uses infrared illumination. Were a large-scale, compulsory iris-scan deployment to be considered, there might be very strong public reaction against the introduction of the technology.

Need for a Proprietary Acquisition Device

Although many biometric technologies require specialized acquisition devices, the lack of competition in the iris-scan market means that the need for a specialized acquisition device could limit the technology's growth. Only a handful of companies manufacture acquisition devices, all under license from Iridian. If these companies are unable to deliver robust components, there are no competitors waiting to capitalize, as in finger-scan and facial-scan. This is not an inherent weakness of the technology, but a reality that deployers must be aware of when implementing the technology.

Iris-Scan: Conclusion

Because the most fundamental argument on behalf of biometrics is increased security, a highly accurate technology such as iris-scan has tremendous appeal. The challenge facing the technology is not to improve its resistance to false matching, but to ensure that its capabilities in real-world environments parallel its capabilities in a laboratory setting.

For a description of the latest iris-scan vendors and technologies, visit www.biometricgroup.com/wiley.

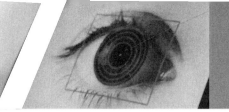

CHAPTER 7

Voice-Scan

Voice-scan technology utilizes the distinctive aspects of the voice to verify the identity of individuals. Voice-scan is occasionally confused with speech recognition, a technology that translates what a user is saying (a process unrelated to authentication). Voice-scan technology, by contrast, verifies the identity of the individual who is speaking. The two technologies are often bundled: Speech recognition is used to translate the spoken word into an account number, and voice-scan verifies the vocal characteristics against those associated with this account.

Voice-scan combines elements of behavioral and physiological biometrics: While the shape of the vocal tract determines to a large degree how a voice sounds, a user's behavior determines what is spoken and in what fashion. Voice-scan technology is text-dependent, meaning that a user must recite a particular phrase or word to be recognized—the system cannot verify a speaker speaking random snippets of text.

Voice-scan's strengths include the following:

➤ It is capable of leveraging telephony infrastructure.
➤ It effectively layers with other processes such as speech recognition and verbal passwords.
➤ It generally lacks the negative perceptions associated with other biometrics.

Voice-scan's weaknesses include the following:

- It is potentially more susceptible to replay attacks than other biometrics.
- Its accuracy is challenged by low-quality capture devices, ambient noise, and so on.
- The success of voice-scan as a PC solution requires users to develop new habits.
- The large size of the template limits the number of potential applications.

Components

Voice-scan systems are similar to facial-scan systems in that they leverage existing acquisition hardware and are based primarily on proprietary software engines that provide template generation and matching. Voice-scan systems are often tuned to operate with certain types of acquisition devices, whether microphones, land telephones, or mobile phones.

A user's spoken phrase is converted from analog to digital format and transmitted to a local or central PC for template generation and storage (during enrollment) or template generation and matching (during verification). For desktop verification applications, the engine that provides template-based functions may reside on a local or a central PC, or may be Web-enabled. In more common telephony applications, the software engine is either located within the institution with which users are interacting (such as a financial institution) or hosted by a third party that offers outsourced biometric verification.

The results of voice-scan comparisons are tied directly into existing authentication systems, often replacing methods such as manual operator verification and last-four-digit routines. The more complex an institution's existing authentication scheme, the more challenging the back-end integration of biometric responses becomes. Voice-scan systems often operate directly in conjunction with speech recognition systems, such that account identification and user verification are achieved through a single process (see Figure 7.1).

How It Works

Voice-scan technology leverages a number of existing processes, particularly when implemented in telephony applications. Aside from enrollment processes, which can be cumbersome, voice-scan is among the biometrics least likely to impact an institution's current interaction with employees or customers.

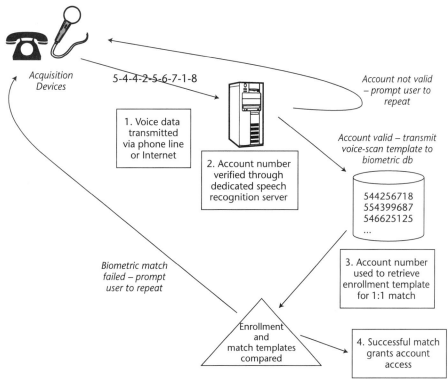

Figure 7.1 Voice-scan and voice recognition—integrated.

Data Acquisition

Voice-scan can utilize any audio capture device, including mobile and land telephones and PC microphones. The performance of voice-scan systems can vary according to the quality of the audio signal as well as variation between enrollment and verification devices, so acquisition normally takes place on a device likely to be used for future verification.

The voice-scan data acquisition process is straightforward. During enrollment, an individual is prompted to select a passphrase or to repeat a sequence of numbers. The passphrases selected should be approximately 1 to 1.5 seconds in length—very short passphrases lack enough identifying data and long passwords have too much, both resulting in reduced accuracy. The individual is generally prompted to repeat the passphrase or number set a handful of times, making the enrollment process somewhat longer than most other biometrics.

In many cases, one or more of the initial utterances is rejected because the user speaks too loudly, speaks too softly, or speaks outside of the acceptable time parameters.

Acquisition is more difficult on PCs than on telephones and is more difficult on mobile phones than on landlines. Most users are still unfamiliar with the paradigm of speaking into a PC microphone—especially in an office environment, many are mildly embarrassed at speaking a password to a PC. Positioning one's self at the proper distance and speaking at the proper time require practice and are normally associated with a number of failed attempts. Employees who use headset microphones and earpieces, such as in call centers, are less likely to encounter this problem. In addition to process flow and interactivity problems with PC acquisition, there is the matter of background noise. Air conditioners, buzzing lights, conversational noise, and copiers are all factors that can disrupt an enrollment or verification.

Telephones are less susceptible to these factors. Users are generally comfortable with the paradigm of prompt and response with a telephone, and voice-scan vendors have developed methods of filtering out noise commonly found on telephone lines. Mobile phone usage, due to reception problems and vocal distortion and echoes, is more likely to result in unusable data being acquired.

Data Processing

Once a sufficient data set has been acquired for enrollment or verification, voice-scan systems process the vocal recordings prior to template creation. This processing includes eliminating gaps at the beginning and end of the recording, so that only the passphrase is used in template generation, and filtering out nonspoken frequencies such as those found in telephone lines and microphones (see Figure 7.2).

Distinctive Features

Voice-scan technology measures a variety of vocal qualities, many of which are not detectable by human listeners. In addition to pitch or fundamental frequency, voice-scan algorithms will often measure other variables such as gain or intensity, short-time spectrum of speech, formant frequencies, linear prediction coefficients, cepstral coefficients, spectograms (time-frequency-energy patterns), and nasal coarticulation. These and other vocal qualities are used in the matching process.

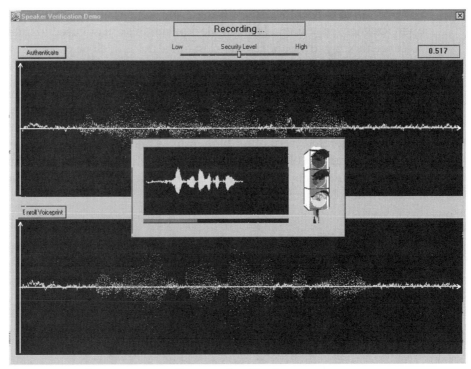

Figure 7.2 Voice patterns: enrollment and verification.

Many of these features used during template creation are replicable only by the human voice, such that they cannot be replicated by high-fidelity recording and playback. While the recording may closely match the spoken phrase, the speakers or monitors used to replay the voice introduce nonhuman elements into the vocal pattern. The possibility of recording and replay attacks is not eliminated, but is significantly reduced.

Template Creation

The overall system of generating a voice-scan template usually relies on a statistics-based pattern matching system, most commonly Hidden Markov models. Hidden Markov models are generalized profiles that are formed through the comparison of multiple samples to find characteristically repeating patterns. During enrollment, template generation relies on the capture of multiple voice samples that are analyzed to determine the qualities that can be relied upon for later recognition. Hidden Markov models are not used exclusively for voice technology; they are also used in other fields such as protein research in order to determine similarities in the chemical structure of different proteins.

Once a generalized profile of the utterances is established, the profile, or voiceprint, is stored within a template for use in later comparisons against a live sample. The size of a voice template tends to be large when compared with templates generated from other biometric technologies. Voice-scan templates can be as large as 10K—small compared to the space that an actual voice recording would occupy, but larger than the 100- to 1,000-byte templates generated in most biometric systems.

Template Matching

Production voice-scan technologies are not capable of one-to-many identification, operating in one-to-one authentication mode. When a user attempts verification, voice-scan systems compare the live submission to the generalized profile and return a statistical rating of the likelihood that the live speaker matches the speaker described by the generalized profile. Due to the inherent variability of speech, as well as the behavioral aspect of the technology, users may unintentionally reduce the system's effectiveness by changing their speech between enrollment and verification. Speaking at a different pitch, speed, volume, or cadence qualitatively changes vocal characteristics and challenges the matching ability of voice-scan algorithms. As in any behavioral biometric system, the performance of the matching system is contingent on users' motivation to verify.

Testing has shown that certain voice-scan technologies are better than many might expect at matching users correctly (limiting false nonmatch rate) while rejecting imposter attempts (limiting false match rate). The template matching process has clearly improved as newer iterations of the technology have emerged.

Deployments

Voice-scan has not been deployed as widely as finger-scan, hand-scan, or facial-scan, although many preconditions of large-scale voice-scan implementations have been met: robust core technology, existing acquisition devices, and need in the marketplace for automated call center authentication. The following implementations indicate the potential range of voice-scan solutions.

Voice-scan is used for postrelease programs to ensure compliance with probation, parole, and home detention terms. While outside of facilities, people can be difficult to track, and such tracking can require extensive manpower and administrative resources. Another challenge is that typical methods of tracking, such as scheduled visits by officers or telephone calls, can result in a long delay between the time that a tracked individual disappears and the authorities become aware that he or she is missing.

Voice-scan is an attractive solution to this problem and has been deployed by T-NETIX and Buytel. In order to verify that an individual is at home while under house arrest, an automatic system telephones them at regular intervals. Using a challenge-response recorded dialogue, the system automatically verifies that the individual is at home through voice-scan. If the voice answering the phone does not match the voice template, or if no one answers the phone, the system immediately alerts officers so that appropriate action may be taken.

A similar curfew enforcement system using voice-scan was adopted by the New York City Department of Corrections. The solution employed by the NY DOC involves the distribution of a pager to the juvenile offender, who is paged regularly after his or her curfew. The offender has a limited amount of time in which to return the phone call, at which point voice-scan is used to verify his or her identity and caller ID is used to ascertain whether the offender is at an authorized location.

Though voice-scan is not widely deployed, pilot projects and isolated deployments are under way. In 1999, Bacob, a Belgian financial institution, became the first European bank to provide a system for customers to make secure transactions by telephone. Keyware implemented the system along with Voxtron, a telephony company. A similar transaction service was implemented for Allied Irish Bank using Buytel's SpeakerKey technology, allowing account access via telephone. SpeakerKey technology has also been used in financial organizations in South Africa and in other parts of the world.

In Australia, Timemac Solutions has used Nuance voice-scan and speech recognition technology to allow stockholders to trade shares via telephone. Also using Nuance technology is Charles Schwab & Co., which began testing a system of telephone-based account access utilizing voice-scan in late 1999.

Voice-Scan Strengths

Voice-scan's strengths are found in the core technology and in its relation to existing processes.

Ability to Leverage Existing Telephony Infrastructure

One of the largest challenges facing large-scale implementation of biometrics is the need to deploy new hardware to employees, customers, and users. Telephony-based voice-scan implementations are capable of circumventing this problem, in particular those implemented in call center and account access applications. Without the addition of hardware at the user end, voice-scan systems can be installed as a subroutine through which calls are routed before access to sensitive information is granted. The ability to use existing telephones means that voice-scan vendors have hundreds of millions of authentication devices available for transactional usage today.

Synergy with Speech Recognition and Verbal Account Authentication

Similarly, voice-scan is able to leverage existing account access and authentication processes, eliminating the need to introduce unwieldy or confusing authentication scenarios. Automated telephone systems utilizing speech recognition are currently ubiquitous due to the savings possible by reducing the amount of employees necessary to operate call centers. Voice-scan and speech recognition can function simultaneously using the same utterance, allowing the technologies to blend seamlessly. Voice-scan can function as a reliable authentication mechanism for automated telephone systems, adding security to automated telephone-based transactions in areas such as financial services and healthcare.

Traditional methods of telephone authentication require a user to provide personal information, which any individual in possession of this information can enter. Voice-scan can then serve as a privacy-enhancing alternative to invasive questioning. When designed as a system to decrease administrative overhead in a call center, voice-scan can increase efficiency even if the false rejection rate is comparatively high. Users who are not authenticated can be routed through standard manual authentication procedures; any users who successfully verify in an automatic verification system decrease the system's administrative overhead. Even if the percentage of users opting in to a system is relatively small, significant administrative effort can be saved.

Resistance to Imposters

Though inconsistent with many users' perceptions, certain voice-scan technologies are highly resistant to imposter attacks, even more so than some finger-scan systems. While false nonmatching can be a common problem, this resistance to false matching means that voice-scan can be used to protect reasonably high-value transactions. This resistance extends beyond imposters' inability to guess the correct passphrase or account number, though this is also a factor that increases system security. Tests in which imposters are given a valid account number to attempt to defeat show substantial resistance to imposter attempts.

Lack of Negative Perceptions Associated with Other Biometrics

Traditionally voice-scan has not been used in law enforcement or tracking applications or in other areas where it could be perceived as a Big Brother technology. Voice-scan thus lacks the associations of finger-scan, facial-scan, and other biometric technologies. There is little fear that voice-scan data can be tracked across databases or used to monitor individual behavior, partly because voice-scan technology is often based on user-selected passwords that vary from application to application. Voice-scan thus effectively skirts one of the largest hurdles facing other biometric technologies—that of perceived invasiveness.

Voice-Scan Weaknesses

A handful of weaknesses relating to performance and perception will likely delimit the areas in which voice-scan can be effectively deployed.

Effect of Acquisition Devices and Ambient Noise on Accuracy

Certain acquisition devices and environments, such as PC microphones in noisy offices or mobile phones with poorer reception, reduce the ability of voice-scan systems to match users. Data unrelated to spoken words, such as static, ambient noise, or echoes, masks the distinctive features used in voice-scan systems. This places telephone-based voice-scan at the mercy of line conditions, in which poor connections can render the system ineffective. Though these issues do not lead to an increased likelihood of false matching, since much of the incoming data is unusable, they do lead to false rejection. Users are then likely to be inconvenienced.

Additionally, testing shows that verification through a different device than the original enrollment device, such as enrollment on a landline telephone and verification on a mobile telephone, can significantly reduce accuracy. In order for voice-scan to provide flexible authentication in a range of environments, these problems must be overcome.

Perception of Low Accuracy

Users unfamiliar with voice-scan technology often assume that it is susceptible to break-in attempts. This reduces user confidence in these systems and may limit the value of transactions users and deployers feel should be secured through voice-scan systems. This perception is attributable to the fact that people recognize one another based on voice, and yet skilled impressions and recordings easily fool people. Impressions do not fool robust voice-scan systems, since imitators copy behavioral components of an individual's voice, not physiological components. The perception of voice-scan as being easily compromised is likely to be a more serious problem for voice-scan than the actual risk of fraud.

Users also often express concern that a voice-scan system will not be able to recognize them if their voice has been modified by sickness, lack of sleep, mood, or other variables. There is some truth to this concern, but more advanced voice-scan systems are designed to accommodate the normal range of changes in individual vocal aspect. This concern is a factor that may dissuade users from relying on a voice-scan system.

Lack of Suitability for Today's PC Usage

Voice-scan vendors promote their products as an alternative for e-commerce, PC login, network access, and all other logical access applications, much in the same way that finger-scan vendors do. However, though PCs usually have built-in microphones, users are generally unaccustomed to speaking to their computers. Self-consciousness and the inability to speak to one's computer in a consistent and replicable fashion make other biometric technologies—those less reliant on performance, such as finger-scan—better desktop solutions.

This weakness may change over time if individuals begin speaking to their computers habitually for other applications, such as speech-to-text applications. Other applications that can help develop the habits that PC users would need for comfort with voice-scan logical access include Internet-based vocal chat, Web meetings, and telephone calls. Until this happens, voice-scan vendors looking for desktop implementations are likely chasing red herrings.

Large Template Size

The templates used for voice-scan are typically larger than those used for most other biometrics. While finger- and iris-scan templates range from smaller than 250 bytes up to 1kilobyte (1K), voice-scan templates are more likely to

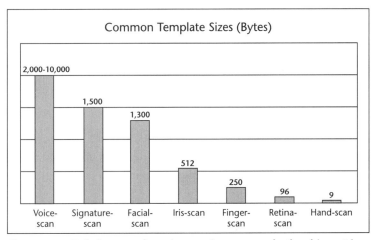

Figure 7.3 Relative template sizes—voice-scan and other biometrics.

occupy 5K, 10K, or more. These file sizes are not large by today's communication and storage standards, but storage on smart cards, memory chips, and other limited-space devices becomes complicated. Template size is likely to become less of an issue as storage size increases and vendors devise methods of reducing the size of voice-scan templates (see Figure 7.3).

Voice-Scan: Conclusion

A technology that most individuals will find easy to use (assuming that they are even aware of its behind-the-scenes operation), voice-scan will be a major biometric solution for a very specific application: telephony-based verification. As telephony applications begin to merge with Web browsing, and PDAs and telephones begin to become one and the same, the potential growth of the solution may increase dramatically.

For a description of the latest voice-scan vendors and technologies, visit www.biometricgroup.com/wiley.

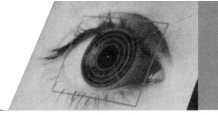

CHAPTER 8

Other Physiological Biometrics

In addition to finger-scan, facial-scan, iris-scan, and voice-scan, deployers should be familiar with a handful of other physiological biometrics, including hand-scan, retina-scan, and automated fingerprint identification systems (AFIS).

Hand-Scan

Hand-scan technology utilizes the distinctive aspects of the hand—in particular, the height and width of the back of the hand and fingers—to verify the identity of individuals. One of the most established biometric technologies, hand-scan has been used for years in thousands of verification deployments. Hand-scan is a more application-specific solution than most biometric technologies, used exclusively for physical access and time and attendance applications.

Hand-scan's strengths include the following:

- It is able to operate in challenging environments.
- It is an established, reliable core technology.
- It is generally perceived as nonintrusive.
- It is based on a relatively stable physiological characteristic.

Hand-scan's weaknesses include the following:

- It has limited accuracy.
- The form factor limits the scope of potential applications.
- The ergonomic design limits usage by certain populations.

Components

All components of a hand-scan system—acquisition hardware, matching software, and storage components—reside within a standalone device. Hand-scan systems are capable of storing thousands of user templates in internal memory and can be used in conjunction with cards that store user IDs for template retrieval. Small processors perform the template creation and matching functions, returning responses through a small display panel. Hand-scan systems are normally integrated into existing access control systems, opening doors upon successful authentication, or into time and attendance systems. Hand-scan deployments that include a large number of doors and employees can be centrally managed, such that enrollment templates from one device can be transmitted to a number of other devices. This prevents users from having to enroll on each device they plan to access (see Figure 8.1).

Figure 8.1 A hand-scan device.

How It Works

Because one vendor is responsible for nearly every hand-scan device currently implemented, there is little variation in how hand-scan devices work. Alternative approaches, such as those that measure the structure of two fingers as opposed to the whole hand, work in a roughly similar fashion but are rarely deployed.

Data Acquisition and Processing

In order to enroll or verify in a hand-scan system, users place their right hand on a recessed, covered metal surface. Five pegs ensure that the hand is positioned correctly, with the fingers slightly separated and the palm resting flat. A series of cameras acquire 3-D images of the back and sides of the hand. Users are required to place their hand on the device three times in order to enroll and once to verify.

Hand-scan devices are integrated units, with image acquisition and processing functions effectively inseparable. The acquisition process for hand-scan is extremely simple and brief—enrollment of all three hand-scan images can be done within 5 seconds. Verification, too, is nearly immediate—once the user's PIN is entered, verification takes less than 1 second. If users misalign their fingers, LEDs indicate the problem area.

Some users cannot provide sufficient hand-scan data to enroll. Arthritic users may not be able to extend their fingers as required or they may not be able to flatten their palm on the device. Users with very small hands may not be able to reach the guidance pegs or spread their fingers widely.

Distinctive Features

Users often mistakenly believe that hand-scan readers use palm prints for verification. While forensic searches can utilize palm prints, hand-scan devices do not. Instead, hand-scan devices measure approximately 90 different features of the hand and fingers. Distinctive features include overall hand and finger width, height, and length; distances between joints; the hand's bone structure; and other elements. Hand-scan devices overlook the tips of the fingers, which can vary in appearance as fingernails grow or are cut. Features acquired by hand-scan systems are not highly distinctive, meaning that the technology cannot be used for identification and is not ideally suited for extremely high security implementations.

Template Generation

The distinctive features of the hand and fingers are extracted from the series of 3-D images and recorded into a very small 9-byte template (most biometric templates are at least 100 bytes, and many approach 1,000 bytes). The small template size allows many user templates to be stored in a very small amount of memory, but is also reflective of the relative lack of distinctiveness of hand-scan data.

Template Matching

Hand-scan systems can be somewhat susceptible to both false matching and false nonmatching, due to the relatively indistinct physiological characteristic. While the hand's bone structure and joints are fairly stable characteristics, under certain circumstances the appearance of the hand can change enough to cause problems. Swelling of the hands can mask the underlying structure, as can various hand injuries. As in most biometric systems, thresholds can be adjusted to allow for higher or lower levels of security or convenience.

Deployments

Hand-scan has been deployed in a number of interesting large-scale applications and is also widely deployed in one- and two-unit implementations for employee-facing applications. The following deployments centered on traveler authentication are representative of hand-scan usage.

In the United States, frequent international travelers can utilize the Immigration and Naturalization Service Passenger Accelerated Service System (INSPASS), permitting them to bypass waiting lines in the airport. Using Recognition Systems hand-scan technology, the system has been in place since 1993. The INS reports that 0.64 percent of international travelers were enrolled in the system in 1999 and projects that 0.70 percent will be enrolled during 2001. Currently, there are 40,000 to 50,000 active users of the kiosk-based system, operational in New York, Miami, Los Angeles, Newark, San Francisco, Washington-Dulles, and U.S. preclearance sites in Vancouver and Toronto. Low-risk U.S. citizens and Canadian citizens can register for CANPASS, an implementation similar to INSPASS, except that CANPASS uses finger-scan technology instead of hand-scan.

In 1998, Ben-Gurion Airport in Tel Aviv, Israel, installed a similar system using hand-scan technology to allow Israeli citizens to circumvent lines when traveling internationally. The implementation was recently expanded due to increased passenger demand, processing 50,000 passengers per month (see Figure 8.2).

Figure 8.2 Kiosk-based hand-scan.

Hand-Scan Strengths

Hand-scan has not changed significantly since its initial commercial introduction, meaning that its strengths and weaknesses are fairly well established.

Ability to Operate in Challenging Environments

Hand-scan devices, constructed out of metal and plastic with few easily damaged components, are designed with challenging environments in mind. This enables hand-scan devices to work outdoors, on construction sites, and in other places where many biometric devices would be inoperable. In addition, the features measured by hand-scan are resistant to day-to-day and environmental changes; temperature, humidity, and other environmental conditions do not affect the ability of a hand-scan device to acquire submissions. This enables hand-scan to operate in a variety of environments that may pose a challenge to finger-scan and other biometric systems. Hand-scan can operate through temperature and moisture extremes and can effectively verify users with dirty hands or those wearing thin latex gloves. Because physical access devices are more likely to be exposed to the elements, this represents a major advantage for hand-scan.

Established, Reliable Core Technology

Hand-scan's core technology has remained unchanged for several years, as opposed to other biometrics whose basic operations are still in developmental stages. As a mature technology that has been successfully deployed to tens of thousands of locations, hand-scan systems operate in a relatively predictable and consistent fashion. This is in contrast to other biometric technologies that are, in theory, much more accurate and convenient but whose effectiveness has not been proven over time or in real-world deployments.

General Perception as Nonintrusive

Users of hand-scan devices usually do not find the devices to be intrusive, threatening, or invasive. This can be attributed in part to comparatively simple enrollment and verification processes. Users do not have to give the system their full attention while attempting to verify, and they can comfortably speak with other people, look away, or otherwise divert their attention during verification. Additionally, hand-scan lacks some of the forensic associations that may impact user perceptions of finger-scan systems. Because the technology is deployed in closed systems with proprietary acquisition devices, the privacy concerns that can hinder other biometric deployments do not affect hand-scan deployments.

Relatively Stable Physiological Characteristic as Basis

Once a person has reached adulthood, the dimensions of his or her hand remain relatively stable throughout the person's life. Aging affects the condition of the skin, but not usually the shape of the hand. Similarly, minor injuries that can render finger-scan inoperable, such as cuts, scratches, burns, and abrasions, do not affect the performance of hand-scan. Other temporal changes, such as weight gain and loss, do not affect hand-scan very strongly. Hand-scan measurements are determined more by the bone structure of the hand than by the flesh. With the exception of arthritis and swelling due to pregnancy or hand injury, the hand is not susceptible to major changes that reduce the technology's ability to match.

Combination of Convenience and Deterrence

In most environments, hand-scan systems will be able to enroll and verify a high percentage of users. The ability to process a very large percentage of users reduces reliance on an alternate authentication mechanism. All biometric deployments require a backup verification system for users who are either unable to enroll or unable to verify consistently; hand-scan is less likely than many other biometrics to cause users to rely on fallback authentication.

The security provided by hand-scan systems often takes the form of deterrence. Since users are unlikely to be rejected frequently, users attempting to break in against other accounts should stand out on an audit trail or transaction record.

Hand-Scan Weaknesses

Hand-scan's weaknesses limit its potential areas of use as well as its ability to provide very high-confidence matches.

Inherently Limited Accuracy

The accuracy of hand-scan is inherently limited both by the general lack of physiological variety expressed in the dimensions of the hand as well as by the relatively small number of features that hand-scan can measure. Although fairly diverse, the size and shape of the human hand are not unique in the same way that fingerprint or iris patterns are. The hand dimensions of any two individuals are normally dissimilar, but in larger populations it is almost certain that various individuals will have very similar hand dimensions. Because of this, hand-scan is only capable of one-to-one verification. Other biometric technologies have a greater potential for security than hand-scan, which is valued for deterrence and convenience as much as it is valued for security. Hand-scan is not an ideal technology for applications in which resistance to imposters is extremely important.

Form Factor That Limits Scope of Potential Applications

In order to accommodate hand placement, as well as to house the camera mechanism necessary for image acquisition, hand-scan devices are comparatively tall, with a large footprint—several inches wide and deep, and nearly a foot tall. This limits hand-scan to physical access and time and attendance applications, and precludes the possibility of hand-scan use in any PC-oriented applications. While manufacturers and distributors of hand-scan devices correctly do not target sales at markets other than those requiring physical access and time and attendance, the potential revenues are limited by the lack of logical access implementations.

Price

Hand-scan, largely a physical access solution, competes with various card technologies, as opposed to other biometrics. With typical costs of $1,500 per unit, hand-scan devices are expensive as compared to most card-reading systems, especially considering that many hand-scan applications include ID cards or badges. Comparable finger-scan systems, while less mature but capable of much higher levels of accuracy, are available at a lower cost with broader functionality. The higher price may be attributable to the lack of competition in the hand-scan market.

Conclusion

Hand-scan is normally deployed as a moderately secure solution for physical access and time and attendance applications—one that provides significant advantages over traditional card systems, including elimination of buddy-punching and deterrence of fraud attempts. The core technology has not been altered nor have new areas of implementation been explored in the past few years, making hand-scan one of the more mundane biometric technologies.

The relative stability of hand-scan technology, in addition to its lack of association with privacy-invasive usage, illustrate how difficult it is to characterize biometrics without referring to a specific technology. Compared to finger-scan and iris-scan—both dynamic, formative, and sensitive technologies—hand-scan almost seems to belong to a different industry. As indicated here, this has its advantages and disadvantages.

Retina-Scan

Retina-scan technology utilizes the distinctive characteristic of the retina—the surface on the back of the eye that processes light entering through the pupil—for identification and verification. Developed in the 1980s, retina-scan is one of the most well-known biometric technologies, but is also one of the least deployed.

Retina-scan devices are used exclusively for physical access applications and are usually used in environments requiring exceptionally high degrees of security and accountability such as high-level government, military, and corrections applications. As of this writing, retina-scan is not a commercially available biometric technology: The original manufacturer is working on an updated device, and newer vendors have not officially released products.

However, because of the technology's history and future potential, it is important for deployers to understand the good and bad of retina-scan technology.

Retina-scan and iris-scan are often mistakenly confused with one another or grouped into a single category referred to as *eye biometrics*. Though they are both very accurate biometric technologies that use features of the eye for identification and verification, the similarities end there: The two technologies differ substantially. They measure different physiological features, the software and algorithm technology is very different, iris- and retina-scan hardware and software are dissimilar, and the situations in which they can be successfully deployed differ.

Retina-scan's strengths include the following:

- It is highly accurate.
- It uses a stable physiological trait.
- It is very difficult to spoof.

Retina-scan's weaknesses include the following:

- It is very difficult to use.
- There is some user discomfort with eye-related technology.
- It has limited applications.

Components

Retina-scan systems are self-contained, with acquisition hardware as well as template processing components within a dedicated device. The results of retina-scan matches are often tied to a door control or other physical access system, with successful matches triggering door release.

How It Works

The following process describes acquisition and processing for the single type of retina-scan technology to be commercialized. This unit is a relatively large, heavy device that rests on a surface or mounts to a wall or doorframe for physical access. The device contains a keypad for input and an embedded lens for image acquisition. Variations on the core technology, such as hand-held units, have been put forth as potential solutions but have not been commercialized (see Figure 8.3).

Figure 8.3 Retina-scan device.

Image Acquisition

Since the retina is small, internal, and difficult to measure without the proprietary hardware and camera systems specifically designed for retina imaging, image acquisition is a very difficult process.

In order for the unit to acquire retina images, the user first positions his or her eye very close to the unit's embedded lens, with the eye socket resting on the sight. Beneath the lens, within the device itself, is an imaging component consisting of a small green light against a white background. The user views this light through the lens; when triggered, the light moves in a tight circle, measuring the retinal patterns through the pupil. In order for a retinal image to be acquired, the user must gaze directly into the lens, remaining perfectly still while focusing on the imaging component. Any movement defeats the image acquisition process and requires that the imaging process be triggered again. A small camera captures an image of the retina through the pupil. The acquisition of a single retina image takes 4 to 5 seconds under ideal conditions.

During enrollment, between three and five acceptable images must be acquired. Since the first one or two images acquired are almost invariably rejected due to excessive movement, the enrollment process can be relatively lengthy. Including the time to trigger the process, respond to system prompts, and acquire sufficient images, enrollments can easily take over 1 minute. Many users cannot enroll at all, even after several minutes. On the other hand, it is possible for highly acclimated users to be identified within 2 to 3 seconds—the identification process is much quicker.

Distinctive Features

The retina's intricate network of blood vessels is a physiological characteristic that remains stable throughout the life of a person. As with fingerprints and iris patterns, genetic factors do not determine the exact pattern of blood vessels in the retina. This allows retina-scan to differentiate between identical twins and provide robust identification (see Figure 8.4). The retina contains at least as much individual data as a fingerprint, but, unlike a fingerprint, is an internal organ and is less susceptible to either intentional or unintentional modification. Certain eye-related medical conditions and diseases, such as cataracts and glaucoma, can render a person unable to use retina-scan technology, as the blood vessels can be obscured.

Template Generation and Matching

Once a device captures a retinal image, the software compiles the unique features of the network of retinal blood vessels into a template. Retina-scan algorithms require a high-quality image and will not let a user enroll or verify until the system is able to capture an image of sufficient quality. The retina template generated by the originator of the technology is a mere 96 bytes, one of the smallest of any biometric technology.

Retina-scan has robust matching capabilities and is typically configured to do one-to-many identification against a database of users. However, because quality image acquisition is so difficult, many attempts are often required to get to the point where a match can take place. While the algorithms themselves are robust, it can be a difficult process to provide sufficient data for matching to take place. In many cases, a user may be falsely rejected because of an inability to provide adequate data to generate a match template.

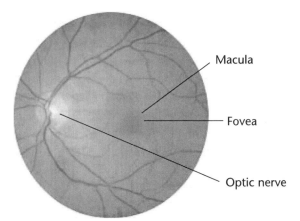

Figure 8.4 Distinctive retinal patterns.

Deployments

Retina-scan technology is best deployed in high-security environments where user convenience is not a priority, particular those involving employees in military, power plant, or sensitive laboratory environments. These deployments will normally involve a single retina-scan unit—a particularly sensitive door may need to be controlled through a biometric, but not an entire facility.

Because of the high degree of effort required to use the technology, it is very difficult to ask citizens or customers to interact with a retina-scan system. However, in the mid-1990s, the state of Illinois actually piloted retina-scan technology as a means of deterring and detecting duplicate welfare recipients. Registration for benefits required that users enroll in a retina-scan system; any users who attempted to enroll for benefits more than once would be detected in the retina-scan database. As would be expected, the technology simply proved too difficult to implement in a compulsory, comprehensive, citizen-facing deployment with poorly trained users.

Retina-Scan Strengths

Retina-scan has two interrelated strengths: resistance to false matching and the stability of the underlying physiological trait.

Resistance to False Matching

Retinal patterns are highly distinctive traits, sufficient to enable 1:N identification. The large amount of data that retina-scan algorithms can leverage, along with the stability of this data, yields a biometric technology highly resistant to false matching. Although the technology is susceptible to other errors due to its difficult operation, well-trained and cooperative users find retina-scan capable of very reliable identification.

Stable Physiological Trait

Unlike some of the other traits used in physiological biometrics, the retina patterns of an individual remain very stable throughout a person's life. Also, the retina is located deep within a person's eye and is thus highly unlikely to be altered by any environmental or temporal conditions, which can render other biometric technologies incapable of operation. Injuries are also highly unlikely to damage the retina, although some diseases can impede retina-scan's operations. For high-security implementations, the likelihood of an imposter creating a fake retina capable of authenticating to a retina-scan device is very low.

Retina-Scan Weaknesses

Existing retina-scan technology has a number of weaknesses that limit its effectiveness for most applications.

Difficult to Use

Retina-scan is more difficult to use than most other biometric technologies, with enrollment requiring prolonged concentration and effort, and identification requiring a well-trained and highly motivated user. Taken by itself, these are not extremely burdensome requirements, but other technologies are much easier to use. Iris-scan, for example, can acquire iris images within 1 to 2 seconds from up to 3 feet away. Finger-scan devices acquire data almost instantly, allowing a user to be verified or identified with modest effort and concentration.

Most individuals are physiologically capable of using retina-scan, although certain conditions can obscure the retina or render its patterns unreadable. Individuals who are blind or have weak vision may be unable to effectively operate a retina-scan device. Also, users with any type of palsy that prevents them from holding their head very still for several seconds at a time will be unable to use today's retina-scan technology.

User Discomfort with Eye-Related Technology

Both iris-scan and retina-scan face the problem of general user discomfort with eye-oriented authentication technologies. This issue is more problematic for retina-scan technology, as the user must position his or her eye very close to the device, an act that can make some users feel uncomfortable. The machinery that drives the circular motion of the unit's imaging components is audible, adding to the perception of invasiveness. Users commonly fear that the device itself or the light inside the device can harm their eyes in some way; other individuals, when introduced to the technology, simply will not use it.

Some discomfort may be attributable to the fact that the inability to use a retina-scan device can be linked to eye diseases. The type of imaging technology used in retina-scan can be affected by medical conditions. Although retina-scan is not deployed for the purpose of diagnosing disease, the association may raise privacy concerns. If implemented in government or occupational situations in which usage is mandated, the technology would likely meet with resistance on the grounds that medical information may be unnecessarily collected while users are being identified.

Limited Applications

The design and operation of retina-scan devices available to this point in the market limit the potential range of deployment. Generally, the only environments in which retina-scan can be effectively deployed are high-security, low-volume physical access and time and attendance operations, in which inconveniencing users is an acceptable cost of heightened security. It would be very difficult to use retina-scan for shift changes or similar applications, due to ingress and egress issues. The devices themselves are relatively large, heavy, and expensive, and are thus not appropriate for usage in most desktop environments.

The difficulty of using the devices also limits the potential populations that the devices may face. Retina-scan users must be highly motivated to be identified successfully while accepting the unit's difficult operation. Unmotivated or reluctant users are likely to lose patience while being identified, which can undermine the effectiveness of the system as a whole. The devices cannot be used effectively with elderly populations or with users who have very poor eyesight.

Conclusion

Retina-scan is not likely to become a widely deployed technology, notwithstanding the common references made to the technology in the media and in popular culture. Other biometrics can provide most if not all of the benefits of the technology with fewer of the drawbacks. In highly specialized applications wherein difficult operation is actually a benefit—such as protecting a weapons facility—retina-scan may be a suitable solution, as potential imposters are unlikely to be capable of even submitting retina-scan data for identification.

Although retina technologies to date have proven comparatively difficult to use and slow to verify, newer versions of the technology may operate faster and more easily, and thus be more suited to different applications. The stumbling blocks for retina-scan are neither the robustness of the technology nor the suitability of the retina for verification. Rather, the form factor and operation of existing devices limits the technology's capabilities; this may change in future versions.

As of this writing, new retina-scan technologies are being developed that claim to be better, cheaper, and faster, with radically different design and operation. If newer retina-scan devices can operate at greater distances and can acquire data more quickly than their predecessors, many new retina-scan applications may be forthcoming.

Noncommercialized Technologies

A handful of noncommercialized biometric technologies have been under development. Table 8.1 is a brief overview of biometrics that may emerge over the next two to four years.

Table 8.1 Noncommercialized Technologies

TECHNOLOGY	CHARACTERISTIC	OUTLOOK
DNA identification	Use of phenotypic DNA material for verification or identification	Very likely to be commercialized for use in public and private sector applications; raises questions of privacy, invasiveness, data misuse
Gait recognition	Rhythmic patterns associated with walking stride	Possible use in surveillance, but many factors to overcome before this approaches viability (clothing, angle of acquisition, need for broad acquisition area)
Nailbed identification	Vertical ridges beneath human fingernail	Speculative
Subcutaneous hand-scan	Tissue structure below surface of skin on the palm	Reasonably strong—prototype units already built, has attracted a fair amount of attention
Thermal facial-scan	Infrared patterns emitted by face below skin	Developed in mid-'90s; biometric characteristic was showed to be highly distinctive, but technology not commercialized due to expense
Vein identification	Vein patterns on back of hand	A physical access product was developed in the late '90s; efforts still under way to commercialize the product as a physical-access solution

Automated Fingerprint Identification Systems (AFIS)

Biometric systems known as *automated fingerprint identification systems* (AFIS) conduct large-scale searches against databases of stored fingerprint templates and images. AFIS technology is capable of identifying a single individual from a database of up to millions of individuals using 1, 2, 4, even 10 of an individual's fingerprints. Though a redundancy, AFIS deployments are commonly referred to as *AFIS systems*.

AFIS should be considered an entirely different type of biometric technology than finger-scan, facial-scan, iris-scan, and all other technologies covered to this point. AFIS has existed for much longer than any other biometric, has enrolled many more users than all other biometrics combined, is a much more mature and proven technology, and operates in strictly defined environments. Most biometric systems are expected to provide verification and identification in seconds; AFIS systems can take minutes, hours, even days to return a match. Whereas most biometrics are emerging technologies, gradually finding applications for which they are well suited, AFIS technology has found its applications and markets. AFIS technology bears a very different relation to privacy than other biometric technologies, as AFIS systems store fingerprint images and are designed entirely to identify individuals from fingerprints. While nongovernment institutions are unlikely to ever deploy AFIS technology, it is important for anyone interested in biometrics to understand the nature and capabilities of AFIS.

Components

AFIS systems comprise front-end live-scan acquisition devices, software on local PCs that process the initial acquisitions, and central PCs that receive, store, and process fingerprint data (see Figure 8.5). The simplest AFIS systems may all be contained within a single facility—a handful of live-scan devices, attached to local PCs, networked to a low-cost, small-scale AFIS capable of comparing up to tens of thousands of records. As AFIS systems grow larger, they may consist of hundreds of acquisition stations throughout a state or country, dedicated networks for transmission of biometric data, and dozens of dedicated PCs costing millions of dollars and executing millions of matches per second. The AFIS back end, which consists of backup storage components, processing components that search fingerprint databases residing in gigabytes of RAM, queuing systems, and other elements, can be viewed as a black box—fingerprints come in, match results go out.

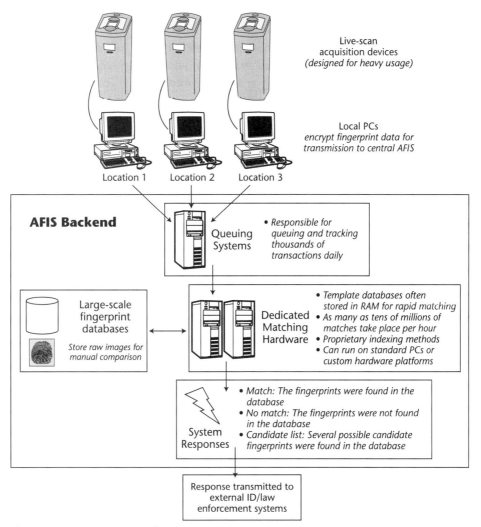

Figure 8.5 Components of an AFIS system.

AFIS systems are designed to search for individuals enrolled in criminal or civil databases, and to transmit the results to external agencies. For example, the FBI's Integrated AFIS (IAFIS), which contains tens of millions of criminal fingerprint records taken from ink-and-roll as well as live-scan images, provides responses to state and local police departments after a criminal search is conducted. In employee background checks, IAFIS transmits match results (whether an applicant has a felony arrest or conviction) to employers directly or through intermediaries.

How It Works

AFIS follows the basic biometric process of image acquisition and processing and template generation and matching, with the added challenge of having to ensure that the data coming in is sufficiently robust to match against millions of users. While AFIS systems utilize templates and proprietary classification methods to perform large-scale searches, they also store fingerprint images to enable manual matching by forensic experts.

Data Acquisition

Enrollment is a critical element of AFIS deployments, as it defines the ability to match against large databases. As the number of fingerprints in an AFIS database increases, so too does the likelihood of false matching. In order for AFIS technology to conduct accurate searches against large databases, high-quality fingerprint images must be acquired during enrollment and identification. High-quality images can be acquired through the traditional ink-and-card method, in which fingerprints are captured on paper and then converted into digital format. Increasingly, jurisdictions collect digital fingerprint images through live-scan devices. These live-scan devices range in size from bulky desktop peripherals to photocopier-sized units, and range in cost from several hundred to tens of thousands of dollars. Live-scan fingerprint technology represents a significant improvement over the traditional ink-and-card method, as fingerprint quality can be immediately assessed and the fingerprint image reacquired if necessary. In employee background checks, a large percentage of ink-and-card fingerprints are rejected by state and federal agencies due to unacceptable image quality.

Systems can require between 1 and 10 fingerprints, depending on the application; criminal applications and employee background searches require 10, welfare and other entitlement programs, because they generally search smaller databases, often require 2. Since high-quality fingerprint images are required to generate robust enrollment templates and execute large searches, enrollment must be supervised. Unsupervised users may be uncooperative or simply may not know how to provide quality data. In criminal searches, the entire print is rolled from nail to nail, so that the maximum amount of data is acquired. This is a much more difficult process than the mere flat placement of fingers (referred to as a *slap*); rolling can take several minutes and requires a highly trained operator. Any sort of distortion in rolled or flat prints increases the likelihood that a user will evade an AFIS search. Supervision is also required to ensure that the correct fingers are placed in the correct order. If an individual surreptitiously places a left middle finger instead of a left index finger, the fingerprint data will be misclassified and the user may be capable of evading detection on subsequent large-scale searches.

Live-Scan: High-Quality Fingerprinting Systems

Finger-scan devices acquire fingerprints for the purpose of template generation and matching in peripheral and standalone systems. These devices are typically inexpensive and small, and are designed for modest daily use. A different type of fingerprinting technology, known as *live-scan,* is used in large-scale fingerprinting applications in which image quality is imperative. Live-scan systems are designed to acquire one, two, or four fingerprints at a time and can also acquire rolled fingerprints when necessary. As opposed to the flat placement found in finger-scan systems, live-scan systems often require that fingerprints be rolled from nail to nail such that the entire print is acquired (see Figure 8.6).

As opposed to finger-scan devices, which may perform functions such as template creation and storage within the unit itself, live-scan devices are only concerned with acquiring and transmitting high-quality fingerprint images. External systems, typically AFIS, perform template generation and matching with live-scan devices as a front end.

Law enforcement, employee background checks, and public-sector benefits programs are typical live-scan applications. These systems replace the traditional ink-and-roll method of fingerprinting, which often generates low-quality, distorted prints. Live-scan systems are built for heavy use with potentially uncooperative individuals and can cost from several hundred dollars to tens of thousands of dollars.

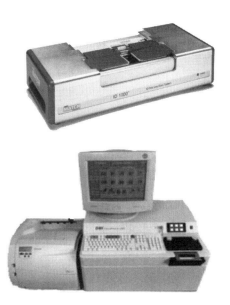

Figure 8.6 Live-scan systems.

Data Processing

In many AFIS systems, quality checks are performed on fingerprint data prior to transmission to the central processing component. This ensures that effective searches can be executed and allows for immediate reprinting of users whose data is insufficient. The types of image processing steps described in Chapter 4, "Finger-Scan"—grayscale processing, thinning, and so on—also take place in most AFIS systems.

Distinctive Features

The average fingerprint has a substantial amount of distinctive data unique to an individual and different from finger to finger. As multiple fingerprints from the same user are collected, the ability to reliably identify users grows exponentially. AFIS systems utilize distinctive features beyond the minutiae and patterns used in finger-scan systems. AFIS vendors have developed proprietary classification schemes and image analysis algorithms in order to increase the distinctiveness of the fingerprint data. These matching processes are the core intellectual property of AFIS vendors and are the results of decades of work in studying and processing fingerprint images.

The use of multiple fingerprints is essential to the success of AFIS deployments, as there is not enough data in a single fingerprint to reliably search against much more than 50,000 to 100,000 fingerprints. However, using multiple fingerprints in the search process does not have a directly multiplicative effect—that is, a system able to match 1 fingerprint against a database of 50,000 fingerprints cannot match two fingerprints against billions of records. Generally speaking, two fingerprint templates can be effectively used on systems of several million, while systems with tens of millions of users require that three to four fingerprints be acquired.

AFIS Matching

AFIS vendors have devised a variety of methods to reduce the challenges of searching large-scale AFIS databases. Though fingerprints are distinctive, once a certain database size is reached it is difficult to search effectively; in addition, low-quality fingerprints are more difficult to search reliably. The primary challenge is to reduce the penetration rate, or the percentage of a database that must be searched for each new enrollee. A penetration rate of 100 percent means that the entire database must be searched; a penetration rate of 10 percent means that only 1/10 of the database must be searched, increasing the likelihood of an accurate response. Some of these methods are very simple, such as filtering the database so that only males are searched when a male enrolls. Others are complex, such as classifying a set of fingerprints according to physiological characteristics and conducting comparisons only against similarly classified prints.

Depending on the requirements of a given deployment, jurisdictions can configure AFIS systems to return several potential matches if multiple fingerprints are found in a large-scale search or to return strong matches. When searching a criminal fingerprint database in order to solve a crime, it may be beneficial to return all possible matches. In this case, the system returns the fingerprint images corresponding to the matching records, at which point a fingerprint expert manually compares the suspect's digital fingerprint images to the sets of fingerprints returned from the AFIS search. Dozens of fingerprints may be returned for examination in this type of search. In public benefits programs, where the potential harm resulting from an unsuccessful search is lower, a jurisdiction may decide to return only strong matches in order to minimize the costs and time involved in resolving duplicate matches. Because of the many variables involved in AFIS searches, standard performance measures such as false match and false nonmatch are difficult to apply; however, the issue of failure-to-enroll is still present because some users are unable to provide sufficient data to enroll in biometric systems.

Large-scale AFIS systems can be quite expensive and complex, requiring immense processing power and storage capabilities. Systems may be required to perform millions of fingerprint matches per second in order to provide an adequate response time. Consider, for example, an AFIS associated with a multiyear project to deploy a trusted national ID card. Such a system might need to accommodate 35 million individuals and 4 fingerprints per individual. In order to enroll a nation's population within three to five years, tens of thousands of citizens would need to be processed by the AFIS every day. As the AFIS reaches 10 million enrolled citizens, the system would need to be capable of performing millions of comparisons for each new enrollee to ensure that he or she has not already enrolled. Once enrollment approaches maximum capacity, the total number of comparisons per day would run into the tens of billions. Furthermore, if the jurisdiction requires responses within a short time, the comparisons must be executed that much more quickly.

Deployments

AFIS is the most widespread application of biometric technology, developed in the mid- to late 1970s to automate the process of searching the tens of millions of ink-based fingerprint cards in FBI files. AFIS technology is used around the world primarily for law enforcement and, to a lesser degree, for employee screening, public benefits programs, and large-scale national ID programs. When arrested suspects are fingerprinted in order to determine their identity, when latent fingerprints are lifted from crime scenes for processing, when financial services professionals are fingerprinted for background searches, and when welfare recipients are fingerprinted in order to detect and deter duplicate enrollments, AFIS technology is at work.

The use of AFIS in civil applications such as public benefits and national ID programs represents a potential growth area for the core technology. The market for criminal searches of arrested suspects and job applicants is fairly well established; however, the potential for nonforensic AFIS databases is largely untapped. The following applications are typical of civil AFIS deployments.

The Republic of the Philippines engaged in a six-year program beginning in 1998 to distribute national social security cards. The program is expected to reach a total of 35 million recipients. The card is designed to prevent fraud and to facilitate secure distribution of benefits to legitimate cardholders, and offers both 1:N and 1:1 transactional functionality. The card combines one- and two-dimensional barcodes as well as a magnetic stripe and other security features. Both Argentina and Nigeria are expected to deploy large-scale national ID systems, using AFIS technology to establish residents' identities.

A number of U.S. states and jurisdictions have deployed AFIS technology to eliminate double-dippers, or individuals receiving welfare benefits under multiple identities. The Arizona Fingerprint Imaging Program (AFIP) was established in 1998 and requires all Arizona welfare recipients to submit to finger-scan when applying for welfare and when receiving benefits in the form of general assistance, needy family assistance, and food stamps. Texas's Lone Star Image System (LSIS) has been operating since early 1998, using AFIS technology to prevent fraud in family assistance and food stamp programs. Los Angeles County implemented AFIS technology for its Automated Fingerprint Image Reporting and Match (AFIRM) system. This system is in the process of being deployed statewide. Since 1995, New York State's Public Assistance program requires that users enroll into an AFIS database to receive benefits. This system is estimated to have prevented millions of dollars in fraud. Connecticut's welfare fraud reduction program, begun in 1996, is designed primarily for 1:N operation but is also capable of 1:1 recipient verification.

How AFIS and Finger-Scan Differ

Although both finger-scan and AFIS technology are based on distinctive characteristics of fingerprints, and both base their matching on templates derived from fingerprints, there are several differences between the two biometric disciplines. These differences extend from the initial acquisition device, which in AFIS systems is a live-scan device as opposed to a finger-scan peripheral, to the back-end matching processes.

- Finger-scan systems normally provide 1:1 authentication, producing match/no-match decisions within seconds. AFIS systems perform large-scale identification with response times ranging from minutes to hours.
- Finger-scan systems normally store templates, not fingerprint images. AFIS systems normally store fingerprint images along with the templates used during large-scale searches.
- Finger-scan systems often store data on local PCs, devices, or smart cards. AFIS systems are predicated on central database storage.
- Finger-scan systems normally use a single finger-scan template for 1:1 matching; additional finger-scan templates are acquired in case of injury or false rejection. Law enforcement AFIS systems utilize up to 10 fingerprints for large-scale 1:N matching, while civil AFIS systems utilize between 1 and 4 fingerprints for large-scale 1:N matching.
- Finger-scan systems utilize data from a flat placement of a fingerprint; most AFIS systems require that the finger be rolled fully from nail to nail on a live-scan device.
- Finger-scan systems utilize inexpensive peripherals designed for moderate usage levels. AFIS systems utilize relatively expensive live-scan device scanners built for heavy usage.
- Finger-scan systems are deployed in public-sector, private-sector, and home applications. AFIS technology is deployed almost entirely in the public sector.

Conclusion

AFIS technology differs in fundamental ways from other biometric technologies and does not normally compete with technologies such as finger-scan, facial-scan, and iris-scan. Instead, AFIS is used in particular types of deployments where it is the only technology that can operate effectively. Because of the familiarity of the fingerprint as a means of identification, as well as the need to identify noncooperative individuals, AFIS technology is likely to continue to be a large component of the biometric industry for years to come.

For a description of the latest hand-scan, retina-scan, and live-scan vendors and technologies, visit www.biometricgroup.com/wiley.

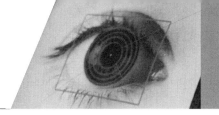

CHAPTER 9

Other Leading Behavioral Biometrics

While voice-scan is currently the most prominent behavioral biometric, signature-scan and keystroke-scan will likely find increasing deployment in a range of environments. This chapter discusses how these behavioral biometric technologies work, as well as their strengths and weaknesses.

Signature-Scan

Signature-scan technology utilizes the distinctive aspects of the signature to verify the identity of individuals. The technology examines the behavioral components of the signature, such as stroke order, speed, and pressure, as opposed to comparing visual images of signatures. The technology has not seen broad usage, but could eventually be widely utilized in electronic document authentication.

Signature-scan's strengths include the following:

➢ It is resistant to imposters.
➢ It leverages existing processes.
➢ It is perceived as noninvasive.
➢ Users can change signatures.

Signature-scan's weaknesses include the following:

- Inconsistent signatures lead to increased error rates.
- Users are unaccustomed to signing on tablets.
- It has limited applications.

Components

Signature-scan systems consist of acquisition hardware, such as pens and tablets, tied into local or central template processing components. While the signature is an extremely common authentication method, signature-scan requires specialized devices to measure behavioral aspects; signature acquisition devices commonly found in retail establishments cannot be used in biometric systems. This acquisition hardware generally consists of electronic signature tablets capable of measuring the speed and pressure of a signature. Less commonly, signature-scan systems may utilize specialized pens capable of measuring these characteristics when signed on standard paper.

The signature, along with the variables present during the signing process, is transmitted to a local PC for template generation. Verification can take place against a local PC or a central PC, depending on the application. In employee-facing signature-scan applications such as purchase order authentication, local processing may be preferred; there may be just a single PC used for such authorization. For customer-facing applications, such as retail or banking authentication, centralized authentication is likely necessary because the user may sign at one of many locations.

The results of signature-scan comparisons must be tied into existing authentication schemes or used as the basis of new authentication procedures. For example, in a transactional authentication scenario, the *authorize transaction* message might be sent after a signature is acquired by a central PC. When signature-scan is integrated into this process, an additional routine requires that the signature characteristics be successfully matched against those on file in order for the *authorize transaction* message to go forward. In other applications, the results of a signature-scan match may simply be noted and appended to a transaction. For example, in document authentication, an unsuccessful comparison may be flagged for future resolution while not halting a transaction. The simplest example would be a signature used for hand-held device login: The successful authentication message merely needs to be integrated into the login module, similarly to a PIN or password.

How It Works

Signature-scan, unlike traditional signature comparison, measures the physical activity of signing. While a system may also leverage a comparison of the visual appearance of a signature, or *static signature*, the primary components of signature-scan are behavioral. There are very few physiological elements in signature scan; theoretically, a person could learn to sign in exactly the same manner as another person. Fortunately, the behavioral characteristics that define signature-scan technology are very difficult to mimic, meaning that it is difficult to break into signature-scan systems.

Data Acquisition

The majority of signature-scan technologies require the writer to sign on an electronic tablet that can record the dynamics of writing. The matching ability of the technology is often dependent on the capabilities of the tablet, and the performance of signature-scan software can be affected by the kind of tablet used for sample acquisition. High-end tablets may have extremely robust pressure sensitivity, able to precisely measure changing pen pressure and detect the position of the pen when lifted from the tablet between writing strokes. Other tablets may not measure pressure at all, only being able to measure the difference between pen down and pen up. In this manner, the functionality of the capture device imposes limits on the variables that can be used, and, hence, the matching capability of the core technology. Acquisition may also take place on stylus-operated PDAs. To the degree that signing will be an important component of PDA usage, this may be an increasingly common implementation of signature-scan technology.

The data acquisition process is similar for all signature-scan vendors. Since the signature is a highly variable characteristic, a series of signatures is required for enrollment. This ensures that inconsistencies from signature to signature are not incorporated within the enrollment template. Users with very long signatures are more likely to encounter problems during enrollment and verification; the abundance of behavioral data present in longer signatures makes it difficult to locate consistently replicable features. In addition, very short signatures are more prone to false acceptance, as there are fewer permutations in the signing process.

Successful system operation requires that users enroll and verify in the same environment. Variables, such as signing from a seated or a standing position, as well as the area around the tablet where a user might rest his or her upper arm during signing, must remain consistent to minimize false nonmatches. (see Figure 9.1).

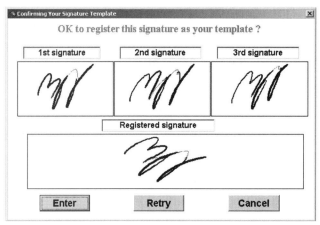

Figure 9.1 Signature-scan interface.

Data Processing

The data acquired during acquisition includes factors such as the speed, stroke order, and pressure of the signature, which combine with appearance to constitute a biometric signature. Not a great deal of processing takes place before distinctive feature location. The duration of a signature may be normalized—extended or shortened to a fixed time in order to facilitate direct comparison. Also, anomalous signature elements such as stray marks outside of the main signature field may be erased. The signature image itself can also be compressed prior to transmission.

Distinctive Features

Traditional signatures are generally regarded as unique, although it is also well understood that a practiced forger can create extremely convincing copies of a given person's signature. The features measured by signature-scan are much harder to forge, since the specific physical behaviors enacted while writing cannot be ascertained from examining a written signature or by observing a person signing. The dynamics measured by signature-scan systems are a conceptually new component of signatures, since such features could not be measured before the existence of electronic writing tablets.

Like other biometric technologies, signature-scan systems extract various features from the raw submission in order to generate a compressed template. Specific details recorded by signature-scan may include the total time taken to sign, the ratio of pen-up to pen-down time, the speed of the strokes, the pressure applied, the number and direction of the strokes, and the total size of the signature, among other variables. The weight given to these measures, as well as other factors that may be utilized, are held as secrets by signature-scan vendors.

Template Generation

Since signature images, even when compressed, are larger than most biometric templates, signature-scan templates do not include images. Instead, numerical values corresponding to the beginning and ending points, inflection points, and relative pressure of each of the signature's components are used to generate templates. These templates can range in size from slightly over 1K to approximately 3K. These templates, large by biometric standards, reflect the variety of data present in a typical signature. Within the signing area, signature data can be present at any point on the X-Y axis and may be written in countless ways. This explains the size of signature-scan templates as well as their positive and negative performance elements.

Template Matching

More so than in most other biometrics, signature-scan technology weighs a number of independent factors when determining whether a match has taken place. An imposter would need to correctly mimic these independent factors in order to be accepted. A user attempting to verify might sign with the identical stroke order and pressure, but with much slower speed, and therefore be rejected.

IBG's Comparative Biometric Testing has shown that signature-scan technology is prone to false nonmatches but is not prone to false matching, meaning that a large amount of data varies from signature to signature. A major challenge for signature-scan vendors is to develop templates that can overcome the day-to-day variations in signatures without being so flexible as to allow imposter verification. Because of the variability in user templates, signature-scan technology can be used only in verification applications and cannot be used to identify individuals.

Deployments

Because of the novelty of the technology, there are few real-world deployments of signature-scan technology outside of standalone PC and PDA authentication for individual users. Charles Schwab & Co. began to allow customers elective use of a signature-scan system for new account applications. Primarily for customer convenience, the system is designed to maximize the efficiency of transactions. In this type of implementation, a user whose signature is verified biometrically may be exempt from more extensive authentication processes, while the likelihood of imposter authentication is very low. This implementation is typical of the direction of the signature-scan industry.

The systems currently deployed to acquire signatures at point of sale and for contractual purposes are not related to biometric authentication, merely to signature capture and storage. The wide acceptance of the signature paradigm suggests that adding a biometric step to this process would not be viewed as invasive, but may provide more security than a standard signature. For this to happen, however, new acquisition devices must be deployed in customer and employee-facing applications.

Signature-Scan Strengths

Signature-scan technology's strengths make it well suited for a specific band of applications in which signatures are part of existing processes and in which nonrepudiation is important.

Resistant to Imposters

The first challenge of a biometric technology has traditionally been to resist imposter attempts. Because of the large amount of data present in a signature-scan template, as well as the difficulty in mimicking the behavior of signing, signature-scan technology is highly resistant to imposter attempts. Any effort to trace the signature, for example, will likely be rejected, as the appearance is not a major factor in determining whether a match has taken place.

As a result of the low false match rate, deployers can be confident that successfully matched users are who they claim to be. Therefore, the technology can be used to ensure that high-value transactions cannot be repudiated. If signature-scan technology is made a precondition of opening a document or executing a trade, deployers can be confident that the transactions can have been executed only by authorized users.

Leverages Existing Processes

A potentially widespread use of signature-scan is anticipated to be in document authentication, as a supplement to the standard signing process. The signature itself is designed to provide a measure of authentication, and the addition of signature-scan can reduce the likelihood of fraud. The most obvious example of this is financial applications such as banking and check cashing, which already utilize signature authentication. The widespread reliance on signatures in many industries and environments provides many potential signature-scan applications.

Signature-scan, in conjunction with PKI, can be used to authenticate electronically transmitted documents. A typical system produces an electronic document embedded with an encrypted packet, enabled by a signature, that both authenticates the signer and verifies that the document has not been altered since signing. Other biometrics can fill the same role, but user experience and comfort with the role of signatures as document authenticators can lead users to prefer signature-scan for document authentication. On paper documents, handwritten signatures have traditionally been used for document authentication, and signature-scan may be viewed as an appropriate biometric for electronic document authentication.

In addition to leveraging existing processes, there may be situations in which signature-scan can leverage existing hardware. Signature capture tablets that provide information on signature characteristics can be used as front ends to signature-scan systems, meaning that a significant impediment to deployment—the lack of acquisition hardware—is overcome.

Perceived as Noninvasive

The handwritten signature is a low-tech authentication method, commonly perceived as useful and nonthreatening. Signature-scan, by extension, is widely perceived as a more secure version of the same noninvasive and nonthreatening process. Users are unlikely to feel concerned that signature-scan could violate their privacy, and the act of signing in a particular fashion is not easily subject to coercion. Moreover, the technology cannot be used for identification against a database of signature templates (in contrast to static signatures, which can be searched with limited effectiveness).

Users Can Change Signatures

When comparing biometrics to other verification methods such as passwords and tokens, users commonly express concern over the unalterable nature of biometric data. Passwords and tokens are temporary and arbitrary, and can easily be replaced, although biometrics are usually inextricably tied to individual identity and can thus never be replaced. Signature-scan is an exception to this category, as a user can voluntarily modify the manner in which he or she signs.

Even though a well-designed physiological biometric system such as finger-scan or facial-scan will be difficult to spoof if biometric data has been compromised, the inability to change physiological data causes some users to mistrust physiological biometrics. While all biometrics are highly resistant to spoofing, users may not feel comfortable using a system for fear that someone could somehow learn to compromise it. Signature-scan allows these users to change their biometric information if the need arises and can thus alleviate their fears despite their mistrust.

Signature-Scan Weaknesses

A handful of weaknesses of signature-scan technology, related to the core technology and to the method by which data is acquired, limit the potential range of deployments and the degree of confidence deployers might have in certain comparison results.

Inconsistent Signatures Lead to Increased Error Rates

As with any biometrics, signature-scan does not work equally well with all users. Individuals who do not sign their names in a consistent fashion, whether out of habit or due to muscle control-related illnesses, may have difficulty enrolling and verifying on signature-scan systems. During enrollment, a series of signatures must be similar enough that the system can locate a large percentage of common characteristics between the enrollment signatures. During verification, enough characteristics must remain common to determine with confidence that the authorized person signed. In addition to the biometrically measurable ways in which the same signature might vary, users often unconsciously vary the signed name itself, including or excluding a title, a nickname, or a middle initial. The result of signature variability may be reflected in higher failure to enroll or false nonmatching.

Testing has shown that the challenge for signature-scan systems does not lie with their ability to reject imposters, but rather with their ability to verify legitimate users, especially over time. This means that while a successful verification is a strong indication that a legitimate user has been authorized, it is unclear whether a failed authentication attempt is an imposter attempt or a legitimate user being incorrectly rejected. Signature-scan systems are best suited for deployments in which users are strongly motivated to match—that is, those in which there is a positive result associated with verification. However, because of the high false nonmatch rate, there cannot be a strong sanction imposed on nonmatches. This is a conundrum for signature-scan vendors:

A low false match rate renders the technology appropriate to protect valuable data or transactions. However, as indicated by the higher false nonmatch rates, a substantial number of legitimate users may not be able to match, and a deployer most likely cannot implement punitive measures such as locking accounts.

Users Unaccustomed to Signing on Tablets

The process of signing on an electronic tablet differs from that of signing on paper. Both the tactile and visual feedback differ, such that users may have difficulty signing in a consistent fashion. When using electronic tablets for the first time, users will often sign in an affected or stylized fashion, increasing the likelihood of inconsistent signatures, false nonmatches, and failure to enroll. As signature-scan technologies that leverage ink and paper emerge, this will become less of a problem.

Limited Applications

While signature-scan is well suited for its target markets, its potential future markets are limited in scope to those in which signatures are currently acquired for authentication. While there are millions of such points of sale where signatures are acquired, this has traditionally been a very difficult market to move into because of the number of legacy systems. Numerous applications for which biometrics are well suited—such as PC login and e-commerce—are very unlikely to utilize signatures.

In environments where signatures are acquired, the handwritten signature is usually trusted without a biometric layer of security. Potential deployers may not perceive the need for the additional overhead (enrollment, integration, security, exception processing) involved in the deployment of a signature-scan system.

Conclusion

Signature-scan is likely to play a modest role in the biometric industry, integrated into applications and processes in which signatures are already acquired and in which nonrepudiation of a transaction is an important element. Assuming that the core technology is improved to the point where failure-to-enroll and false nonmatch rates are reduced, signature-scan could grow along with similar technologies such as handwriting recognition and signature capture.

Layered Biometrics

Layered biometric solutions are those that require the submission of more than one biometric characteristic for verification, such as finger-scan and voice-scan or finger-scan and facial-scan. These solutions are seen as a method of reducing false match and false nonmatch rates, as an imposter is unlikely to defeat two biometrics, and a user unable to verify on one technology may be able to verify on another. Although a handful of vendors are capable of implementing layered biometric solutions, the percentage of real-world biometric implementations that actually require verification through more than one biometric is very small.

There are two types of layered biometrics: *parallel* and *serial*. Parallel layered solutions require that the user submit multiple biometrics in a single authentication procedure. For example, one company's biometric solution requires that the user speak a password while the facial appearance, lip movement, and voice pattern are compared against previously enrolled users. The results are weighed and a pass/fail decision rendered. Serial layered solutions combine the result of two or three separate authentication processes to authenticate users. Serial layered user interaction might consist of verification on finger scan, then facial-scan, then voice-scan. A handful of vendors offer serial solutions.

Layered biometrics are likely to remain niche solutions because they introduce a number of operational complexities into what is intended to be a simplifying process. Having to verify on two biometrics in sequence can be time-consuming and requires that users learn and be mindful of two authentication processes. Furthermore, there is doubt about whether layered solutions provide higher security than a strong biometric alone. Some voice-scan solutions, for example, are subject to high false match rates. When combined with a technology with low false match rates, such as finger-scan, the weaker technology may reduce the layered system's overall capabilities. However, for certain implementations, layered biometrics may be a feasible solution (see Figure 9.2).

Keystroke-Scan

Keystroke-scan technology utilizes a person's distinctive typing patterns for verification. Using normal computer keyboards, keystroke-scan measures variables such as the length of time a user holds down each key and the time elapsed between keystrokes. These behavioral characteristics are somewhat distinctive, are difficult to observe, and are present for any individual capable of pressing keys on a keyboard.

Other Leading Behavioral Biometrics 133

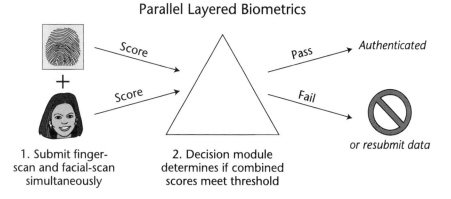

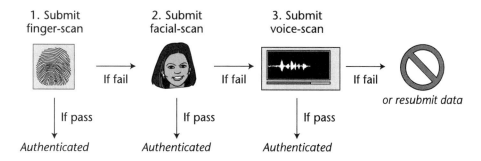

Note: users may need to pass 2/3 or 3/3 biometrics to be authenticated

Figure 9.2 Layered biometrics.

Keystroke-scan is a purely behavioral biometric—without a user actively typing on a keyboard, there is no characteristic to measure. Keystroke-scan is normally deployed in conjunction with passwords and is not currently implemented to monitor users typing on the fly. While there are limitations on the potential accuracy of keystroke-scan technology, it can operate in environments and applications where other biometrics face impediments.

Keystroke-scan's strengths include the following:

➢ It leverages existing hardware.
➢ It leverages a common authentication process.
➢ The password can be changed as necessary.

Keystroke-scan's weaknesses include the following:

- The technology is still in the formative stage.
- It adds only security, not convenience.
- It retains many flaws inherent to password-based systems.

Components

Keystroke-scan technology is a pure software solution, leveraging keyboards (and, in theory, keypads) to enable biometric measurement and authentication. Software installed on a user's PC measures keyboard usage, generating enrollment or verification templates as necessary. These templates can be compared against locally stored enrollment templates or against templates stored on a central server, depending on an enterprise's requirements for security and convenience.

Match decisions from keystroke-scan systems are normally integrated into an application's or an operating system's authentication subsystem. The biometric match occurs before the password is transmitted to the existing application. If a user matches successfully, the password is transmitted as normal. If a user does not match, he or she is prompted to reenter the username/password combination, or the PC may be locked while an administrator is notified.

How It Works

Only one company widely distributes a keystroke-scan product, though a handful of smaller companies have developed variations on the core technology. The following processes reflect the technology currently available in the market, though it is likely that competing technologies will follow similar methods.

Data Acquisition and Processing

Standard keyboards function as acquisition devices in keystroke-scan systems. While standard applications are only concerned with what keys a user types, not the speed or manner of typing, keystroke-scan leverages typing characteristics not utilized in day-to-day computing. Operating systems can measure the duration for which a key is depressed and the time between keystrokes; keystroke-scan systems gather this data and, after a sequence of characters is entered, a distinctive set of characteristics emerges.

Keystroke-scan enrollment and verification are text-dependent; that is, the user must choose a specific word or phrase (username and password) to enroll and verify. In order to collect sufficient data, both the username and password should be at least eight characters long. Enrollment in keystroke-scan can be a cumbersome process, as users are required to type their username and password approximately 15 times. This enrollment can take place in one sitting or it can occur over multiple days with sequential logins, depending on whether convenience or security is an institution's first priority. The advantage of single-sitting enrollment is that the system security is in place from day 1 onward; however, enrollment quality can be questionable. A user's typing patterns when repeating 15 username-password combinations may differ from the patterns present when typing the combination only once. Distributed enrollment is likely to result in more robust system operation, as the data acquired in authentication attempts spread out over a period of time better represents the variability of a user's typing habits.

Data acquisition for keystroke-scan verification is essentially invisible to the user. The only difference between keystroke-scan-enabled passwords and standard passwords is that keystroke-scan will not work if users make corrections when typing the username or password. If a user presses the backspace key in the middle of a username or password, for example, the software requires that the user retype the word or phrase from the beginning.

Distinctive Features

The first question users ask about keystroke-scan technology is whether typing patterns—how long keys are held down, duration between keying instances—differ enough to be distinctive. Testing indicates that typing patterns do differ substantially from user to user, such that keystroke-scan features are somewhat distinctive. The problem is that many users' patterns differ from login to login, meaning that an individual's biometric keystroke data is not consistent. Similar to signatures, keystroke patterns seem to change from iteration to iteration for many users and can also change perceptibly over time.

Touch typists are most likely to have consistently replicable keystroke-scan data. However, it remains to be determined whether touch typists share common typing traits that make it difficult to tell them apart. Deeply ingrained typing habits may be shared among strong typists to a greater degree than among mediocre typists, especially on commonly typed words and phrases such as passwords.

Template Generation and Matching

Templates generated in keystroke-scan systems store both username and password data, as well as the behavioral data related to typing patterns. As security levels are increased or decreased, the required degree of correlation between the enrollment and verification templates is adjusted accordingly. For example, a low-security implementation may allow for moderate variation in a few of the typed characters. A high-security implementation would require similarity across all characters between enrollment and verification. When verifying through keystroke-scan, users are authenticated sequentially first by the biometric system, then by the existing password schema. The benefits and drawbacks of existing password systems remain.

Keystroke-Scan Strengths

Keystroke-scan has a number of unique qualities that separate it from traditional biometric technologies.

Leverages Existing Hardware

A major impediment to desktop deployments is the need to deploy and support proprietary acquisition devices, such as finger-scan devices, or alternatively to configure and train users on peripherals such as cameras and microphones. Keystroke-scan, by leveraging a component available on most every desktop in the world, circumvents the hardware problem entirely. Being a software solution, it can be deployed to a desktop environment with less effort than most other biometrics. Issues of hardware upgrades, available ports, and ergonomics are largely absent from keystroke-scan deployments. Additionally, in network environments, an acquisition device exists at every PC, allowing users to authenticate from any workstation.

Leverages Common Authentication Process

One of the main difficulties in deploying biometric systems is introducing new processes into an established process flow. PC and network login is almost invariably protected by passwords, and virtually everyone is familiar with standard password procedures. Deploying finger-scan, for example, requires that users be trained on new and unfamiliar devices and processes. Keystroke-scan users continue to utilize existing authentication methods and do not normally need any additional training or processes. The only new process might be an increased awareness of the manner in which they type their usernames and passwords.

By leveraging an existing authentication process, keystroke-scan provides an additional degree of security. While other biometrics may provide increased convenience or ease of use, keystroke-scan does not; keystroke-scan deployments are most likely driven by a need for incrementally greater security than that offered by existing password-based systems. Even if installers of a keystroke-scan system do not trust the technology to effectively block every invalid login attempt, a keystroke-scan system can only make it harder for imposters to get into a system, never easier. A keystroke-scan system that rejects 75 percent of imposter attempts—by normal biometrics standards, a terrible rate—is still providing a 75 percent reduction in imposters compared to standard password authentication.

Can Enroll and Verify Users with Little Effort

Enrollment on a keystroke-scan system can occur transparently to the user. Systems can be configured such that a series of successful logins is recorded, and an enrollment template is generated after a sufficient period of time. This allows a system to be installed and set up without interrupting normal work flow or negatively impacting productivity. Verification similarly lacks additional procedures—users merely need to type their usernames and passwords while the biometric functions transparently. The benefits of transparent enrollment can be substantial, because the enrollment process is one of the more challenging logistical elements of deploying a large-scale biometric system.

Usernames and Passwords Can Be Changed

Potential biometrics users often express concern over the permanent nature of biometric data and the inability of users to change biometric data should it be compromised. In the event that a password and biometric template are compromised, a user of keystroke-scan can reenroll using a new password, rendering the compromised information useless. Templates created by keystroke-scan systems are specific to a particular password and therefore can only perform one-to-one verification. Furthermore, a keystroke-scan system cannot use an enrollment to match a user typing anything beyond his or her username and password combination: The technology cannot surreptitiously identify users typing documents, emails, and so on. Keystroke-scan is unlikely to evoke the privacy concerns associated with many physiological biometrics. Users are less likely to feel that keystroke data is personal data, and they are less likely to view such systems as invasive.

Keystroke-Scan Weaknesses

Along with these compelling advantages, keystroke-scan does face many challenges before it approaches the level of acceptance of mature biometric technologies.

Young and Unproven Technology

The technology behind keystroke-scan is still developmental and has not been subject to rigorous real-world implementation. While the concept behind the technology appears to be theoretically sound, the development of algorithms capable of reliably verifying authorized users while rejecting imposters is by no means a fait accompli. Testing has shown that keystroke-scan systems are susceptible to rejecting authorized users and to authenticating imposters. Whereas many mature biometric systems will reject well over 99 percent of imposters, even at low security levels, keystroke-scan is one of few biometric technologies susceptible to double-digit false matches.

More problematically, the technology is still susceptible to rejecting authorized users at certain security levels. Because there is effectively no fallback procedure for users falsely rejected by a keystroke-scan system, it is essential that the technology be capable of reliably verifying users without rejecting large percentages. The marginal security increase provided by keystroke-scan systems, potentially a significant benefit, is useless if false rejection rates are substantial.

Does Not Increase User Convenience

Most biometric deployments, especially for logical access, are motivated in part by a need for user convenience and a need for greater security. Keystroke-scan is only capable of adding some degree of security and is likely to be less convenient for users than traditional password authentication. The increased incidence of false rejection and the need to be aware of one's typing patterns make keystroke-scan less convenient than the password systems that biometrics are meant to replace.

Retains Many Flaws of Password-Based Systems

Advocates of biometrics for logical access identify many flaws inherent to password-based systems related to security, convenience, and efficiency. Keystroke-scan retains most of the flaws of password-based systems, such that administrative effort is likely to be increased, not decreased. Users are prone to forgetting passwords, which may prevent legitimate users from gaining access. Unless integrated into single sign-on packages, keystroke-scan does not alleviate the need for users to remember multiple passwords for different applications and resources. Security policies that dictate password changes are likely to be kept in place, even with keystroke-scan implemented.

In addition, keystroke-scan usernames and passwords are recommended to exceed a minimum length in order to provide adequate data; although this is good security policy, convenience is again reduced. The end result is increased administrative overhead: In a keystroke-scan system, all of the existing password management processes remain, with the addition of reenrollment for certain users and fallback procedures for falsely rejected users.

Addressing these issues requires a more robust, proven keystroke-scan technology than currently exists. If the biometric component of keystroke-scan can be relied upon to verify nearly all users consistently and can be proven to be sufficiently robust to reject imposters, some of these password management procedures could become superfluous.

Conclusion

While not yet deployed in large-scale operational environments, keystroke-scan has unique and compelling potential as a low-cost, low-impact logical access solution. As opposed to other biometrics, where the distinctiveness of the underlying biometric trait is established, it remains questionable just how distinctive keystroke-scan actually is. If algorithms can be developed that overcome the normal variations in users' typing patterns while resisting imposter attempts, keystroke-scan may become very widely deployed.

For a description of the latest signature-scan and keystroke-scan vendors and technologies, visit www.biometricgroup.com/wiley.

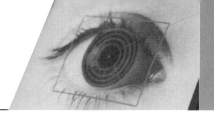

PART THREE

Biometric Applications and Markets

Just as leading biometric technologies differ in fundamental ways, the major biometric *applications* differ substantially in terms of security and convenience requirements, process flow of enrollment and verification, and system design. Deploying biometrics as a surveillance tool, for example, differs substantially from deploying biometrics in PC and network security. In order to deploy biometrics effectively, you'll need to understand the dynamics of your specific application.

In addition to the variety of applications of today's biometric technology, there are various *vertical markets* in which biometrics are deployed effectively. These markets share a common need for biometric identification and verification. However, the roles of biometrics in areas such as health care and financial services—to name just two leading vertical markets—can differ substantially.

Portions of the analyses in Part Three are derived from International Biometric Group's Biometric Market Report 2000-2005, an independent market survey and analysis of the biometric industry.

CHAPTER 10

Categorizing Biometric Applications

While many people are interested in how biometrics will be deployed in a specific market—such as financial services, travel and immigration, or the government sector—the *application* for which a biometric technology is used is a more important distinction. Biometrics exist because of the technology's ability to provide verification and identification in areas such as PC and network access, criminal identification, telephony, physical access, and e-commerce. Understanding biometrics according to the use or application for which a technology is deployed—often referred to as a *horizontal* approach—provides a great deal of insight into how to deploy biometrics effectively. A horizontal approach identifies the problems that biometrics will solve and the benefits they will provide over the next several years.

An approach predicated on *what the biometric is doing* presents a more fundamental distinction than which biometric technology is being used or in which vertical market the technology is being deployed. This point is often overlooked by industry observers intent on portraying the market solely in terms of technology growth or vertical markets. A horizontal approach underscores the substantial differences between biometric applications and allows for a better understanding of issues essential to the biometrics industry, such as security, privacy, and accuracy.

Defining the Seven Biometric Applications

Traditionally, biometric applications are divided into just three categories:

- Applications in which biometrics provide *logical access* to data or information.
- Applications in which biometrics provide *physical access* to tangible materials or to controlled areas.
- Applications in which biometrics identify or verify the identity of an individual from a database or token.

While these categories do indicate the most basic differences among those in which biometrics can be deployed, they fail to capture a number of application-specific factors critical to understanding biometrics:

- In what manner does the individual interact with the biometric system? Is the user supervised?
- Does the user claim an identity before interacting with the system?
- What are the application's requirements for accuracy, enrollment, and response time?
- Is the user motivated to comply with the biometric system? What sanctions are in place for misuse?
- What is the value of the data or materials that the biometric is protecting?
- With what nonbiometric technologies do biometrics compete?

After considering how biometric applications differ so substantially according to these basic criteria, seven horizontal classifications take shape:

Criminal identification. Criminal Identification is the use of biometric technologies to identify or verify the identity of a suspect, detainee, or individual in a law enforcement application. The primary role of the biometric is to identify an individual in order to proceed with, or halt, a law enforcement process. Few technologies can effectively compete with biometrics in this horizontal; without biometrics, it might be impossible to identify a suspect.

Retail/ATM/point of sale. Retail/ATM/point of sale is the use of biometrics to identify or verify the identity of individuals conducting in-person transactions for goods or services. The biometric is used to complement or replace authentication mechanisms such as presenting cards and photo identification, entering a PIN, or signing one's name.

E-commerce/telephony. E-commerce/telephony is the use of biometrics to identify or verify the identity of individuals conducting remote transactions for goods or services. The biometric is used to complement or replace authentication mechanisms such as passwords, PINs, and challenge-and-response interaction. Although there are some differences between e-commerce and telephony applications, most notably in the acquisition device, the two have more in common than you might realize. Both applications involve remote user authentication, and both often require transactional verification. In addition, both applications normally feature unsupervised enrollment and verification processes. This category becomes more important as biometrics are used more frequently to authorize various types of transactions.

PC/network access. PC/network access is the use of biometrics to identify or verify the identity of individuals accessing PCs, PDAs, networks, applications, and other PC resources. The biometric is used to complement or replace authentication mechanisms such as passwords and tokens. Of the seven horizontal classifications, PC/network access is closest to traditional logical access, as previously mentioned. Unlike e-commerce/telephony, PC/network access authentication is not used to authenticate a specific transaction, but instead to grant access to a resource.

Physical access/time and attendance. Physical access/time and attendance is the use of biometrics to identify or verify the identity of individuals entering or leaving an area, typically a building or room, at a given time. The biometric is used to complement or replace authentication mechanisms such as keys, tokens, and badges. Time and attendance is frequently deployed in conjunction with physical access. While these are two separate applications, physical access and time and attendance are linked because they pertain to restricting, registering, or controlling the presence of an individual within a given space.

Citizen identification. Citizen identification is the use of biometrics to identify or verify the identity of individuals in their interaction with government agencies for the purposes of card issuance, voting, immigration, social services, or employment background checks. The biometric is used to complement or replace authentication methods such as document provision, signatures, or vouchers. The biometric may provide unique functionality if it is used to prevent duplicate registration for a public benefit or service.

Surveillance. Surveillance is the use of biometrics to identify or verify the identity of individuals present in a given space or area. The biometric is used to complement or replace authentication methods such as manual monitoring of cameras. Surveillance differs from physical access/time and attendance inasmuch as surveillance does not assume user compliance.

Surveillance also has dramatically different requirements in terms of accuracy and enrollment and verification processes.

As shown in the comparative growth rate chart in Figure 10.1, five applications can be classified as *emerging*—PC/network access, e-commerce and telephony, physical access, retail/ATM/point of sale, and surveillance—while criminal identification and citizen identification are classified as *mature*.

The following chapters will address these seven applications, describing how biometrics are used in each application, the strengths and weaknesses of biometrics in each application, and what deployers need to know to implement biometrics successfully.

In our discussion of biometric applications, we will also classify applications according to a little-used but critical high-level categorization: the capacity in which an individual uses a biometric system.

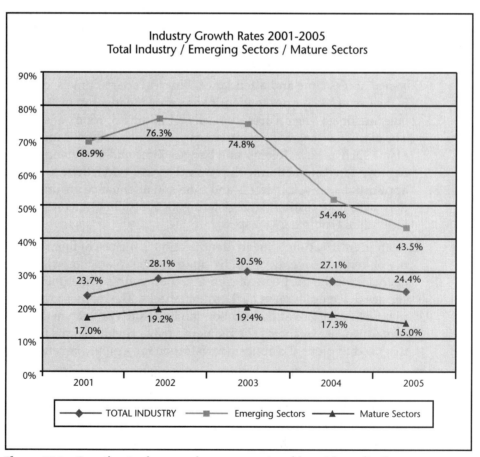

Figure 10.1 Growth rates for emerging versus mature biometric applications.

Capacities in Which Individuals Use Biometric Systems

Assessing biometric applications according to the capacity in which the individual interacts with the biometric system provides a highly revealing look at growth in the biometrics industry. The role a person occupies when interacting with a biometric system is an often overlooked but essential determinant of such factors as privacy, accuracy, and performance.

For example, the potential for most types of privacy infringement in applications wherein the government is authenticating individuals is much larger than in applications in which companies are authenticating customers. In terms of accuracy, systems used to authenticate employees normally must have a higher enrollment rate than systems that authenticate consumers, since the potential risks attributable to workarounds are greater in employee-facing applications. Falsely rejecting a customer will likely have more severe repercussions for a merchant than falsely rejecting an employee.

The three primary roles an individual can occupy when interacting with an authenticating agent are employee, citizen, and consumer. Employee authentication refers to authentication of an individual during the course of employment, in which the individual provides biometric data in order to access information or interact with his or her employer. Accessing enterprise networks and providing fingerprint data for employment-related background searches are classified as employee-oriented revenue; executing a B2B transaction is classified as a consumer role, since the dynamic between authenticating entity and individual differs. Citizen authentication refers to authentication of an individual by a government body for purposes of law enforcement, benefits disbursement, obtaining a driver's license, and voter ID. Consumer authentication refers to authentication of an individual in the course of, or as a prelude to, executing a transaction for goods or services. The seller or trusted agent is one half of the transaction; the consumer is the other.

Not every biometric usage falls within this categorization. A biometric door lock installed by an individual on his or her front door, for example, would be neither citizen, employee, nor customer authentication, because there is no external authenticating agent. However, every important biometric installation and deployment can be classified accordingly.

Introduction to IBG's Biometric Solution Matrix

Biometric technology is used to solve authentication problems—without an authentication problem, biometrics are merely an interesting technology in search of an application. The suitability of biometrics for a given application depends on their ability to address a deployer's specific authentication needs.

The Biometric Solution Matrix is a guide to deploying biometrics for specific applications, designed to help deployers assess the nature of their specific authentication problem. The Biometric Solution Matrix defines the five elements that deployers should consider when deciding whether to implement biometrics: urgency, scope, effectiveness, exclusivity, and receptiveness (see Figure 10.2).

How Urgent Is the Authentication Problem that Biometrics Are Solving?

Biometrics can be used to address mission-critical authentication problems or minor authentication problems. An authentication problem may be deemed urgent as the result of substantial risk to valuable data, assets, revenues, or public safety. Less urgent problems would entail less risk to these elements. Biometric deployments become more important to an institution when the authentication problem they must solve is urgent. In the Biometric Solution Matrix, *urgency* is rated on a scale from 1 to 10.

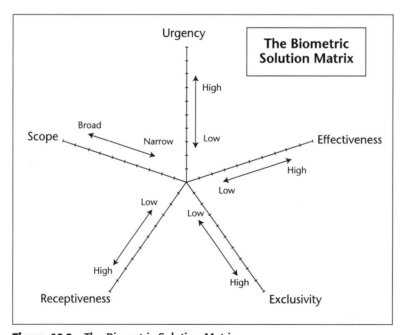

Figure 10.2 The Biometric Solution Matrix.

What Is the Scope of the Authentication Problem that Biometrics Are Solving?

Biometrics can be used to address an authentication problem that is limited in scope, such that only a small percentage of individuals might interact with the biometric system or to address authentication problems encountered by a large number of individuals on a regular basis. Biometrics are more likely to be a strong solution when addressing authentication problems that are broad in scope. In the Biometric Solution Matrix, *scope* is rated on a scale from 1 to 10.

How Well Can Biometrics Solve the Authentication Problem?

The use of biometrics might eliminate an authentication problem or scarcely address it. Biometric technology, when deployed correctly, can solve problems through verification and identification or through simple deterrence, but there are situations in which neither of these approaches is effective. Biometrics are more valuable to deployers when the methods used are highly capable of effectively solving authentication problems. In the Biometric Solution Matrix, the ability to solve a problem, or *effectiveness*, is rated on a scale from 1 to 10.

Are Biometrics the Only Possible Authentication Solution?

Biometrics might be the only solution to an authentication problem or one of many potential solutions. Biometrics are stronger solutions in applications in which they are the only viable alternative to an authentication problem. In the Biometric Solution Matrix, *exclusivity* is rated on a scale from 1 to 10.

How Receptive Are Users to Biometrics as an Authentication Solution?

Biometrics might be welcomed as a necessary authentication solution or they may be dismissed as a possible solution for a variety of reasons. Without receptive and informed customers, employees, and citizens, the potential for biometrics to be an authentication solution is limited. *Receptiveness* is rated on a scale from 1 to 10. By plotting the five elements of the Biometric Solutions Matrix, potential deployers can assess the suitability of biometric technology for their particular application. Applications in which there is significant potential for biometrics will be plotted toward the outside of the matrix; applications in which the potential for biometrics is limited will be plotted toward the inside of the matrix (see Figures 10.3 and 10.4).

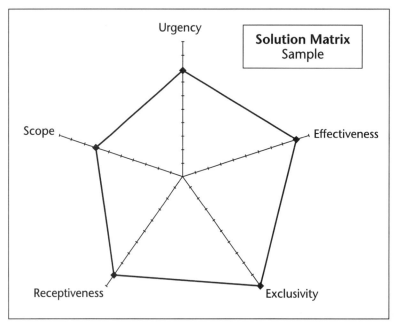

Figure 10.3 Biometric Solution Matrix: strong application-specific solution.

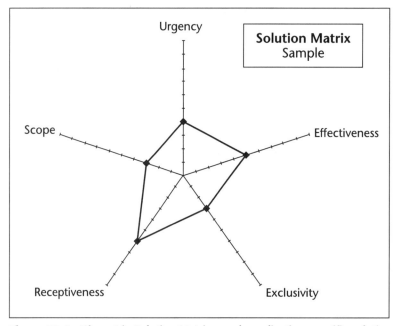

Figure 10.4 Biometric Solution Matrix: weak application-specific solution.

CHAPTER 11

Citizen-Facing Applications

Citizen-facing applications include criminal identification, citizen identification, and surveillance. The defining element of citizen-facing applications is that a government body, normally a state or federal agency, provides authentication and enforces compliance with the biometric system's match decisions. Citizen authentication includes authentication for the purposes of law enforcement, benefits disbursement, obtaining a driver's license, and voter ID.

By their nature, citizen-facing applications are more likely to be mandatory than other biometric applications. In addition, citizen-facing applications are likely to be predicated on centralized storage of biometric data, either at a state or at the federal level. These applications often utilize identification as opposed to verification, as there may be a compelling need to identify a non-compliant subject or to ensure that a subject is not claiming a fraudulent identity. Finally, these applications are generally large-scale deployments, enrolling and authenticating hundreds of thousands, if not millions, of people over an extended period of time. As addressed in Chapter 15, "Assessing the Privacy Risks of Biometrics," many of these characteristics have the potential to be used in a privacy-invasive fashion. When reviewing the following citizen-facing applications, bear in mind how such systems might be deployed in a privacy-sympathetic fashion.

Criminal Identification

Criminal identification is the use of biometric technologies to identify or verify the identity of a suspect, detainee, or individual in a law enforcement application. The primary role of the biometric is to identify an individual in order to conduct law enforcement functions.

Today's Criminal Identification Applications

Criminal identification was the first widespread use of biometric technology, deployed for decades in nonautomated applications. Over the past 25 years, automated fingerprint searches against local, state, and national databases, as well as automated processing of mug shots, have become pervasive criminal identification applications, used around the world.

Future Criminal Identification Trends

Growth in the criminal identification market will be driven by the advent of inexpensive, functional solutions that expand biometric capabilities into new jurisdictions and environments. Price is a major factor in criminal identification systems, since live-scan acquisition systems can cost tens of thousands of dollars. As these systems are incorporated into smaller form factors, they will become more affordable to local agencies. These solutions will include wireless applications capable of transmitting subject data from remote locations to centralized databases to receive real-time, field-based identification. The availability of face databases makes facial-scan an attractive solution for jurisdictions looking to perform 1:N identification on detained individuals, although the accuracy one would expect in an AFIS system will be lacking.

Similarly, less expensive systems will enable the growth of criminal identification solutions in international markets. Whereas there is a degree of saturation in the U.S. markets in terms of criminal identification hardware and software, markets outside the United States are much less saturated. The development of Internet-based, shared local fingerprint and facial image databases will also drive growth, as jurisdictions are able to get faster responses from regional databases than from large statewide or national databases.

The advent of automated DNA solutions could have a tremendous impact on the criminal identification market. Databases with criminals' DNA are already being populated and searched in the United States. Not only is the population of these databases a basic precondition for large-scale searching based on DNA

> ## Is DNA a Biometric?
>
> In its current state, DNA matching differs from standard biometrics in a handful of ways:
> - DNA requires a tangible physical sample as opposed to an image, recording, or impression of a behavioral or physiological characteristic.
> - DNA matching is not done in real time, and currently not all stages of comparison are automated.
> - DNA matching does not employ templates or feature extraction, but rather represents the comparison of actual samples.
>
> Regardless of these basic differences, it is very likely that DNA matching technology will develop to the point that it does become automated and will be capable of performing identification from elements such as naturally occurring oils on the skin. At this point, it becomes a semantic question as to whether a technology that requires a tangible sample can be a biometric. Realistically, DNA matching should be classified as a biometric inasmuch as it is the use of a physiological characteristic to verify or determine identity.
>
> Whether DNA matching will be employed beyond its current use in forensic applications is uncertain. Intelligent discussion on how, when, and where DNA should and should not be used, who will control the data, and how it should be stored is necessary; the conditions under which its usage, collection, storage, and disposal are acceptable must be defined and enforced. These definitions will vary by application, as is always the case with biometric technology. It is illogical to suggest that the use of DNA in public benefits programs should be viewed as an equivalent to the use of DNA in a criminal investigation. Thinking about the role of DNA as a biometric is helpful because it underscores the range of applications for the technology—if used correctly, DNA can be an extremely powerful tool for determining guilt or innocence. On the other hand, DNA matching could expand into areas in which misuse of personal information deducible from specific DNA becomes possible.

samples, but efforts are under way to streamline and automate the matching process.

The primary impediment to growth in the criminal identification sector is its very success. Many jurisdictions have already purchased or deployed criminal identification systems, such that migrating to newer systems may not be warranted (assuming that doing so involves replacement of current systems). This affects live-scan/AFIS solutions much more than facial-scan systems; the market for facial-scan law enforcement systems is largely untapped.

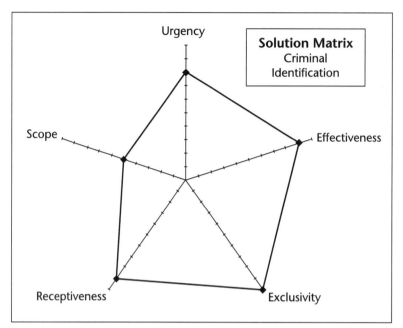

Figure 11.1 Criminal Identification Solution Matrix.

Related Biometric Technologies and Vertical Markets

AFIS technology, with live-scan devices as a front end, is far and away the most frequently deployed criminal identification technology. The fingerprint is a universally recognized method of identification, and the FBI has hundreds of millions of fingerprints on file for conducting criminal identification searches.

Beyond AFIS, facial-scan is used in criminal identification, but with a much lower degree of accuracy. The large databases of facial images collected during the booking process enable searching for close matches, but the technology does not currently provide high-confidence 1:N matching. Over time, it is possible that iris-scan will find limited use in criminal identification applications, as it is able to provide rapid identification by means of a static biometric characteristic. However, the lack of any iris-scan databases limits the use of the technology in this environment.

Clearly, criminal identification applications are the primary subset of the law enforcement market. Other applications can also be classified as law enforcement, such as surveillance and physical access, but all criminal identification applications can be classified as occurring within the law enforcement market.

Criminal Identification: Biometric Solution Matrix

As indicated in the Solution Matrix, various factors make biometrics a strong solution for criminal identification applications (see Figure 11.1). The only limiting factor is scope, meaning that criminal identification is limited to specific environments.

Exclusivity (10/10). Biometrics are the only technology capable of addressing the large-scale identification needs in this market. Individuals claiming false identities or maintaining multiple identities can be identified only through their biometric data. Without automated biometric identification systems, the task of identifying criminals is exceptionally difficult. This explains the maturity of the criminal identification application as compared to other biometric applications.

Effectiveness (9/10). Though not error-free, biometrics have proven to be an effective solution for criminal identification. Biometrics—in particular, AFIS solutions fronted by live-scan devices—can achieve a high level of accuracy and rapid response times, and can be used to accurately identify the vast majority of individuals. The only limitation on effectiveness is the lack of environments in which the technology can currently be successfully deployed. Mobile devices are becoming available in this market but are not fully mature solutions.

Receptiveness (9/10). There is little resistance to the use of biometrics in criminal identification; it is taken as a matter of course that fingerprinting is a necessary process for identification of noncooperative individuals. Normally, receptiveness is a function of individuals' opinions on the technology from a privacy and a cost-benefit perspective. Biometrics, in this application, are seen as a necessary technology overriding privacy concerns and warranting the expense of deployment. The use of DNA in criminal identification has also not met with significant objections, although many issues with regard to collection of DNA from individuals suspected but not convicted of crimes have not been resolved.

Urgency (8/10). Biometric use in criminal identification applications is not always critical, as the individual may have already been adequately identified and only some percentage of individuals are held under fraudulent identities. However, there are many situations in which the ability to perform rapid searches is essential, since an identity match may be necessary to keep a suspect in custody.

Scope (5/10). Most individuals will never be on the receiving side of a criminal identification process, so the scope of the solution is not very broad—it exists within a circumscribed set of law enforcement applications. Although it is true that society in general derives benefits from the functionality that biometrics provide in this area, most individuals will not directly interact with biometrics on a daily basis in a criminal identification environment.

Cost to Deploy Biometrics in Criminal Identification

The costs of deploying biometrics in criminal identification occur through purchase, integration, and servicing of hardware and software solutions, including:

- Acquisition hardware such as fingerprint imaging systems
- Card conversion systems to convert ink-based fingerprint cards to electronic fingerprint databases
- Fingerprint matching systems, including hardware and software
- Facial-scan software capable of indexing and processing facial images
- Deployment, integration, and maintenance of biometric systems

Sales of criminal identification systems are normally executed in multiyear contracts. Although there are off-the-shelf criminal identification products suitable for local jurisdictions, they are normally deployed as part of a long-term engagement. As deployment environments grow larger—to city, state, and regional systems—criminal identification systems grow much more complex and expensive. At one extreme, the FBI's multiyear program to develop its Integrated AFIS—making its tens of millions of fingerprint cards electronically searchable—cost $640 million. Oklahoma engaged Printrak to deploy a statewide identification system, including palm and AFIS capabilities, for $5 million.

Conclusion

As suggested throughout this section, criminal identification is a comparatively mature biometric application. Whereas some basic stumbling blocks impede the deployment of biometrics in areas such as PC/network access and e-commerce—attributable to the novelty of and unfamiliarity with biometrics in these environments—the processes necessary to implement biometrics successfully in criminal identification have been defined. Biometric solutions are normally implemented by the professional services arm of a biometrics firm, responsible for installation, training, configuration, and occasionally even operation of these systems.

In deploying newer biometric systems such as facial-scan and mobile finger-scan solutions, the primary concern of deployers should be long-term viability of the solution. While standards have been developed that define, for example, which live-scan readers are qualified to acquire AFIS images, similar standards have not been developed for newer criminal identification solutions.

Citizen Identification

Citizen identification is the use of biometrics to identify or verify the identity of individuals in their interaction with government agencies for the purposes of card issuance, voting, immigration, social services, or employment background checks. The biometric is used to complement or replace authentication methods such as document provision, signatures, or vouchers.

Typical Applications

A wide range of interactions between individuals and governments is captured under the citizen identification horizontal.

Voting and Voter Registration

Mexico has licensed facial-scan technology to check voter rolls for duplicates in its next national elections. Because large databases of facial images are already available in this type of application, the inclusion of facial-scan technology requires neither the introduction of complex process flows nor the introduction of a new image capture infrastructure. A similar system was implemented in Uganda's 2001 elections, a deployment that met with controversy due to its association with the tactics of Uganda's leadership.

Application for and Receipt of Government Entitlements or Benefits

A number of U.S. states—New York, California, Arizona, Connecticut, and Texas, among others—have made finger imaging a requirement of registration for welfare and other types of public aid. Many of these multimillion-dollar programs were implemented in the early to mid-1990s, and are designed to detect and deter duplicate benefits recipients. These systems have, by and large, been deemed successful. Since many were implemented as part of larger programs aimed at reducing system abuse, the precise savings attributable to the biometric are undetermined. However, states' general willingness to renew and expand these systems may be seen as evidence of their value. Internationally, citizen identification programs designed to streamline or legitimize government benefits issuance are in place in South Africa, in the Philippines, and in various locations in Latin America. Each program has enrolled millions of citizens and is expected to grow into the tens of millions.

Immigration-Related Activities

Biometric technology is used to facilitate border crossing for citizens, for passport issuance and processing, and to verify the identity of refugees. Of the various applications of biometrics in citizen identification, immigration may have the highest growth potential, as countries have a compelling interest in controlling movement across their borders, and citizens have an expectation that processing will be rapid. Examples of this usage include implementation of biometrics to facilitate customs clearing in international airports in the United States, Canada, and Israel; Guatemala's passport registration program, which eliminates duplicate issuance and facilitates lost passport replacement; and the Netherlands' program by which asylum seekers are verified against card-based biometric data.

Driver's License or Identification Card Issuance

Biometrics are used in license and identification issuance programs to ensure that duplicate identities are not created and to enable transactional functionality. Dozens of jurisdictions have implemented or plan to implement these identification programs, including Illinois, Georgia, and West Virginia in the United States; Argentina, El Salvador, Panama, Bolivia, Argentina, and Nigeria; and states in India and China.

Background Checks for Criminal Activity and Citizenship Status

In the United States, citizen identification systems are used to ensure that individuals applying for employment in certain fields have not committed crimes. Though this could be classified as criminal identification, biometric identification is not prompted by a criminal act, but by a legislative requirement placed on employment applications. If a match occurs, what results is not arrest or prosecution, but refusal of employment.

Activities such as an individual purchasing goods through a government Web site or a state employee logging onto a network would not be considered citizen identification. Instead, this category is meant to include authentication events in which an individual's status as a citizen or resident is central to the interaction between the individual and the government.

Future Trends in Citizen Identification

Advances in citizen identification will be driven by the various situations in which strong authentication of citizens is a necessity. Certain elements of citizen identification are more likely to grow in the United States, Canada, and

Europe, while others will find greater acceptance in Asia, South America, and developing countries, where privacy concerns are less commonplace.

A primary growth driver of biometrics in citizen identification will be the need to provide strong authentication in government benefits programs. Most developing countries have some sort of state-sponsored or government benefits issuance system; at the same time, many countries lack the ability to verify their citizens. The use of biometrics to provide government services, because it provides benefits to both government and citizen, is expected to drive biometric revenue growth. The South Africa model, in which biometrics are used to facilitate benefits issuance and prevent government abuse, is extensible to a range of countries.

Issuance of multifunction cards, capable of carrying information such as employment status, emergency medical information, and citizenship status, will also drive the use of biometrics in citizen identification. Developing countries, in an effort to improve services and strengthen the government's ability to interact with its citizens, are exploring the use of cards to enable various services. Because of the transactional elements involved, as well as potentially sensitive data, the use of a biometric to secure access to cards will increase. A resistance to card-based identification programs may limit the use of biometrics in this environment in Western countries, although the balance between desire for security and desire for privacy can shift according to events well outside the scope of the biometric industry.

Voter registration will also be a driver for citizen identification deployments, since authentication is a central part of the voting process. A major question, however, is whether biometrics will be used to facilitate and improve voting processes or as a tool to selectively dissuade voters. This use of biometrics as a citizen identification solution bears close watching.

The availability of high-capacity AFIS systems is a key component of growth in this sector. AFIS technology is the only technology realistically capable of providing the $1:N$ identification necessary to establish exclusive identities. Improvements in the technology have decreased search time and increased accuracy, such that systems will become increasingly deployable in challenging environments.

It is possible that much of the growth in citizen identification systems will occur outside of the United States, Canada, Europe, and Australia. These regions have shown a general suspicion of, and occasional hostility to, government-centered biometric usage. Since many citizen identification applications are meaningful only if data is stored in a database for the purposes of $1:N$ matching, the potential for abuse or misuse is present. These privacy concerns are not shared as strongly outside of Western democracies, although it cannot be ruled out that resistance to government-centered systems will spread to Asia and the developing world.

The logistics of enrollment, card issuance, and identification may limit the deployment of certain types of citizen identification applications. Large-scale national ID projects, even without biometrics, are massive, multiyear undertakings. When biometrics are included, these systems grow more complex and expensive. It is possible that the return on biometrics will not be viewed as warranting the expenses involved.

Citizen Identification: Biometric Solution Matrix

As shown in the Solutions Matrix (see Figure 11.2), biometrics have a great deal of potential as a citizen identification solutions. The only limiting factor seems to be urgency—in some citizen identification applications, the biometrics would provide substantial benefit, but are not an imperative.

Exclusivity (9/10). Biometrics are the only technology capable of performing many of the functions required in citizen identification. Establishing a singular identity, for example, requires the ability to search biometric databases for duplicates. Much of the deterrent effects upon which these systems are based are also predicated on the use of biometrics.

Effectiveness (8/10). Biometrics operate well in citizen identification environments, with limited exceptions. AFIS technology has been shown capable of locating duplicate enrollments in large databases, and facial-scan can be used to effect gross searches. There is also generally some amount of sanction against system misuse, such that efforts to attempt to subvert the biometric are dissuaded.

Receptiveness (7/10). Receptiveness is a major factor in citizen identification systems, because the explicit or implied consent of the governed is necessary to implement large-scale systems. In the United States, Canada, Europe, and Australia, the public has been generally receptive to the use of biometrics, but only to solve specific problems or for circumscribed uses. In other areas of the world, biometrics are more generally accepted without circumscribed uses.

Urgency (6/10). The primary factor limiting biometrics as a solution to citizen identification is urgency. The technology is performing in this environment; the question is whether it is necessary to implement biometrics, or whether the problems involved in citizen identification are not extremely pressing. A further question is to what degree is it desirable or necessary for governments to robustly identify their citizens, and for what purposes.

Scope (9/10). Citizen identification has the potential to be a very wide-ranging solution, not being limited to certain markets. One factor limiting the scope of biometrics in this area is simply frequency of use. In some applications, biometrics might be used to enroll an entire citizenry, but then used only every few years for reauthentication.

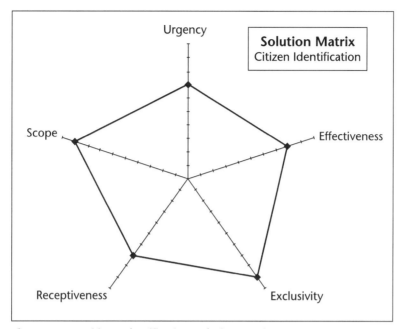

Figure 11.2 Citizen Identification Solution Matrix.

Related Biometric Technologies and Vertical Markets

The biometric technologies most likely to be used in citizen identification are AFIS, facial-scan, and finger-scan. The first two technologies are most capable of performing large-scale 1:N identification, with varying levels of accuracy, cost, and speed. Finger-scan is most likely to be used in transactional verification, expected to be a small part of overall citizen identification.

There is a great deal of overlap between citizen identification and government-sector usage, although the government will use biometrics in a variety of areas outside the scope of citizen identification.

Cost to Deploy Biometrics in Citizen Identification

Most revenues in citizen identification are in the design and deployment of large-scale systems. These projects are multiyear undertakings, in most cases, and involve enrolling, processing, and storing data for millions of citizens. Hardware and software sales may be only a portion of the revenues that go to biometric companies—there are often substantial service and maintenance costs as well. Although many of these revenues go directly to biometric companies, large-scale integrators play major roles in the implementation of many of these systems.

Issues Involved in Deployment

Certain types of citizen identification applications, especially those involving the use of 1:N identification for large-scale card issuance or registration, may be the most difficult biometric systems to design and deploy successfully. While the nature of these systems varies, the main challenges include logistics of enrollment, scalability, performance and error rates, and ensuring individual privacy.

Logistics of Enrollment

Citizen identification systems may reach millions, even tens of millions, of individuals. With the exception of facial-scan systems, which might leverage existing databases of facial images, the process of enrolling these millions of individuals can be extremely complex. A person's identity must be established with a high degree of certainty before enrolling in the system, to ensure a firm foundation for later verification or identification attempts. Enrollment quality must be high in order to ensure reliable long-term verification and identification; this requires supervision by trained staff able to instruct users in proper enrollment methods. The proper data must also be submitted—in a two-finger identification system, subjects may be able to avoid detection by submitting prints in the wrong order. The sheer scale of the enrollment effort—a jurisdiction must enroll over 15,000 users per day over the course of several years in order to populate larger citizen ID systems—is such that hundreds of distributed enrollment stations, tied to a central processing station, may be necessary. A jurisdiction considering deployment of a citizen ID system may need to spend several months evaluating the enrollment process before entering into more technical areas of system design. These issues are in addition to the further logistical challenges of renewal and reacquisition of biometric data at multiyear intervals.

Scalability

For citizen ID systems performing 1:N identification, the question of scalability is paramount. Systems must be capable of anticipating and accommodating population growth, or else effective matching may become untenable. Biometric technologies capable of large-scale operation do eventually reach a point where they can no longer reliably identify individuals in 1:N searches. For example, single-fingerprint searches cannot be reliably conducted on databases well over 100,000 (depending on the specific vendor); even two-fingerprint searches begin to reach maximum database sizes in the several million or tens of millions of users. A jurisdiction that designs a biometric system to scale to 10 million users and comes to be faced with 12 million enrollees over the course of several years will likely need to revisit its core system capabilities.

Response Times

In 1:N systems, an implementation issue closely related to scalability is response time. Response time is a function of the size of the database against which searches are conducted, the number of users being enrolled in a given period of time, the peak enrollment times a jurisdiction encounters, and the back-end processing power. Response times in 1:N systems can be critical: Subsequent processes such as issuing cards or triggering benefits dispersal can be contingent on the results of a 1:N search. Any delays in this process will not only impact a single user but may cause a large backlog. In order to speed response times in large-scale searches, biometric data can be classified or filtered according to biometric categorizations (such as fingerprint patterns) or nonbiometric categorizations (such as gender).

Error Rates

Deployers must ensure that the error rates within their citizen identification systems fall within acceptable bounds. For applications such as benefits registration, the failure-to-enroll rate must be kept as low as possible; individuals unable to enroll cannot be detected in subsequent duplicate identity searches. System false match and false nonmatch rates are critical determinants of system success. The likelihood of an individual registering twice in a system—the false nonmatch rate—must be kept low but cannot be eliminated. In some cases a given user may have a 5 percent chance of evading detection by the system, such that 1 in 20 fraud attempts is successful. Though such a metric would be unacceptable in PC/network access, increasing a system's ability to detect multiple identity fraud by 95 percent while deterring an unknown number of would-be fraud attempts will in many cases be deemed a success. The false match rate may be the most critical metric, as citizens processed differently because of suspected duplicate registration may call the system's effectiveness into question. From a public relations perspective, false matches in citizen identification searches can be highly problematic.

Legacy Systems

The integration or interfacing of citizen ID systems with existing identification systems or databases of public records can also be a major undertaking. Legacy systems may be highly proprietary and incapable of the automated response necessary to enable large-scale identification. Though the biometric functionality of large-scale systems is largely standalone and is not impacted by existing systems, the match/no-match decisions may need to be communicated to a series of external systems.

Privacy-Sympathetic Data Handling

The potential for misuse of biometric data is substantial in citizen identification applications. These systems are generally centralized and may contain a significant percentage of a country's population. In addition, the collection of personal information such as employment identification numbers may be necessary. Deployers must ensure that systems are designed such that even internal systems operators are unable to gain access to biometric data for unauthorized purposes. This may entail both logical and physical separations and protections, such as encryption, when feasible.

Conclusion

Citizen identification is one of the most challenging but potentially beneficial applications of biometric technology. In applications where a country requires a high degree of identity certainty regarding visitors, travelers, benefits recipients, voters, or other segments of its population that interact with government agencies, biometrics are the only effective tool available. Many of the limitations on the potential of biometric citizen identification systems are related more to the logistics of system operation than to the capabilities of the core technologies.

Surveillance

Surveillance is the use of biometrics to identify or verify the identity of individuals present in a given space or area. The biometric is used to complement or replace authentication methods such as manual monitoring of cameras.

Today's Surveillance Applications

Biometric surveillance systems are deployed in the majority of the major casinos in North America and in a handful of police applications to search passersby or event attendees against hit lists of known criminals. Of these two applications, casino usage is much more widespread and less controversial.

The use of automated matching through surveillance cameras has proven controversial in the United States. However, following the events of September 11, 2001, the interest in biometric surveillance increased dramatically, especially for air travel applications. While there were a small number of surveillance applications in airports prior to the attacks, public- and private-sector officials subsequently began contemplating large-scale implementation of the technology to monitor crowds for wanted terrorists.

Future Trends in Surveillance

Depending on the results of these first few very high-profile surveillance implementations, surveillance may come to be a widely deployed technology. It could be argued that one of the primary impediments to other biometric applications—the absence of acquisition hardware—does not apply to surveillance, since millions of security cameras are already installed around the world. Biometric surveillance technology may eventually prove capable of leveraging this infrastructure, installed as a back-end software solution and matching faces that come into camera view against databases of known criminals. The theoretical size of the market for this technology in public- and private-sector applications is extremely large.

Law enforcement officials have already expressed interest in using this technology more broadly, and it is likely that there will be many environments in which the public safety argument will outweigh privacy invasiveness. Security firms will also include automated surveillance capabilities as a part of their product and service offerings, packaged with installation and monitoring of high-end security systems.

Two factors may inhibit the emergence of surveillance applications: performance and privacy. Facial-scan technology, while currently the only biometric capable of operating in surveillance mode, is not idealized for use in surveillance applications. The angle, distance, lighting, and temporal changes introduced between initial image capture (mug shot) and subsequent identification attempts reduce accuracy. There is also the strong possibility that users will be falsely matched, which calls into question both the effectiveness of the system—if thousands of users are falsely matched for each person found on a watch list, it will be difficult to maintain vigilance—and whether this is an infringement on individual privacy. Though surveillance systems may be deployed for their deterrent effect, some efficacy will need to be demonstrated for deterrence to matter.

Surveillance: Biometric Solution Matrix

While some factors underscore the potential for biometrics in this sector, there are some imposing limitations on biometrics as a surveillance solution. Receptiveness, effectiveness, and urgency are all problem areas. (see Figure 11.3.)

Exclusivity (10/10). Biometrics are the only technology capable of performing automated identification in surveillance applications. It is this exclusivity that may drive the growth of biometrics as a solution in this area—biometrics represent a new, automated approach to security not otherwise possible.

Effectiveness (4/10). The effectiveness of biometrics in this area can only be measured in terms of deterrence, because detection is likely to be a rare occurrence. The deterrent effect could prevent criminals, terrorists, or other individuals on a watch list from frequenting areas where the technology is activated. However, due to the limitations of today's facial-scan solutions in identifying subjects in surveillance applications, as well as the extremely small number of individuals on watch lists as compared to legitimate travelers, the odds are against surveillance technology working effectively in this arena. By definition, surveillance is a noncooperative identification solution, because users under surveillance are not making an identity claim and are generally unaware of system operation.

Receptiveness (7/10). Prior to September 11, 2001, biometric surveillance had not been very well-received by the general public. However, since then the balance has clearly shifted toward a preference for security tools and away from an insistence on privacy. To the degree that decisions on surveillance deployments will be impacted by public sentiment—as will be the case in public-sector deployments—the public's newfound receptiveness will increase the frequency of biometric deployments.

Urgency (8/10). As is the case with receptiveness, the events of September 11, 2001, cast an entirely new light on the urgency of biometric surveillance. Prior implementations designed to catch card counters and pickpockets, though not without purpose, came to seem almost inconsequential compared to applications such as air travel security and terrorist watch lists. Whereas previously it was not certain that the problems that surveillance systems attempt to solve—identifying wanted criminals in specific locations or displacing certain criminal elements—was compelling enough to warrant deployment, this concern is no longer expressed.

Scope (8/10). Surveillance solutions have the potential to reach a large number of environments and applications. The potential for biometric applications is heightened because of the range of existing systems that surveillance technology may be able to leverage and the range of application environments.

Citizen-Facing Applications

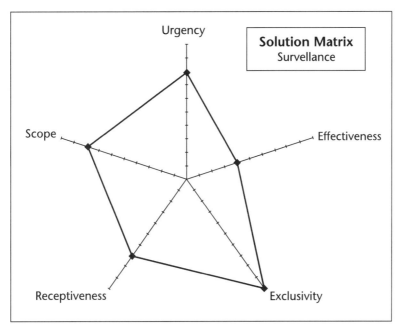

Figure 11.3 Surveillance Solution Matrix.

Related Biometric Technologies and Vertical Markets

Facial-scan is the only technology currently capable of performing in surveillance applications. It is possible that voice-scan could be used to monitor conversations, but only for 1:1 verification—this would not be surveillance in the 1:N sense. Law enforcement and gaming are two markets that have already embraced surveillance, as the result of synergies between these markets and surveillance technology.

Cost to Deploy Biometrics in Surveillance

Costs for surveillance systems can vary widely, depending on whether facial-scan technology can leverage existing devices or requires the installation of new devices for improved performance. The licensing of facial-scan technology is the primary cost involved in surveillance applications. Licenses can be issued for use in specific locations or specific times, or to search against fixed databases. Broader licenses, such as those involved in incorporating facial-scan into a security product offering, may generate larger revenue. Because of the relative novelty of this biometric solution, the prices facial-scan companies can command for the licensing of their software are undetermined. Additional costs are normally borne in additional security staffing and support, as the results of facial-scan searches must be monitored in order to resolve matches and alert appropriate authorities.

Issues Involved in Deployment

Biometric technology is less proven in surveillance applications than in any other horizontal, making the issues involved in deployment even more important to successful system building.

Existing Acquisition Hardware

It is commonly thought that facial-scan technology is capable of effectively leveraging images taken from existing equipment such as CCTV cameras. Unfortunately, due to factors such as distance, angle, lighting, and resolution, such is rarely the case. Effective surveillance requires high-quality cameras positioned at a distance and angle consistent with existing user enrollments. This means that deployers may need to implement dedicated cameras at very specific positions in a building or terminal in order to attain a sufficient level of accuracy.

Size and Quality of Target Database

The effectiveness of facial-scan solutions is heavily determined by the quality of enrollment. In most facial-scan systems, enrollment requires the acquisition of multiple facial images from slightly varying angles, such that the system can accommodate a slight range of movement. In surveillance scenarios, enrollment often occurs through a single low-quality photograph. There is often insufficient data in this photograph to reliably extract identification data, and therefore such photographs are difficult to match. Deployers must make a realistic determination regarding the quality of enrollment images.

Similarly, deployers must assess the expected ratio of surveilled individuals to individuals on watch lists. If, for example, a surveillance system must scan 1 million faces per year in order to identify 10 terrorists—most of whom will not set foot in surveilled locations—the overall efficacy of the system may be questioned. On the other hand, attempting to identify 100 terrorists out of a surveilled group of 100,000 users may be more viable.

Process for Intervention

Surveillance systems, like any biometric system, provide confidence levels in biometric matches, not 100 percent matches. In order to determine whether a flagged match is indeed a match, manual intervention is necessary. Once a person has confirmed that a match occurs, a process must be devised by which the individual is intercepted, bearing in mind that the individual may, even after manual intervention, have been misidentified.

Deterrence versus Detection

Hidden facial-scan systems are unlikely to provide a great deal of deterrence, as some percentage of users will be unaware of their operations. Cameras in clear view, in addition to probably being more effective at identifying individuals, are more likely to elicit what will appear to be odd behavior—intentional efforts to mask one's face or alter facial aspect when faced with a surveillance camera. In many cases, the deterrent element of surveillance technology can be emphasized through intelligent deployment.

Conclusion

Surveillance is a very challenging biometric application, one that arguably matches the least-effective enrollment process with the most-difficult acquisition process. While surveillance is often viewed as a prototypical biometric application, such that the specter of surveillance applications can raise significant privacy concerns, the current state of facial-scan technology is such that surveillance has not yet proven to be an effective use of biometric technology. Improvement in the ability to enroll, acquire, and identify faces from low-quality images and acquisition devices would significantly increase the viability of this solution.

CHAPTER 12

Employee-Facing Applications

Employee-facing applications include PC/network security and physical access/time and attendance. The defining element of citizen-facing applications is that an institution, be it in the public or the private sector, provides authentication and enforces compliance with a biometric system's match decisions. Employee-facing applications are generally closed systems, incorporating a department, a division, or an entire institution's staff.

These applications are also likely to be mandatory and may or may not be predicated on centralized verification. Employee-facing applications are frequently based on verifying a claimed identity, not on identifying employees—the individual's identity has normally already been established. An individual acting in the capacity of an employee is more coercible than a customer, and the privacy implications of this coercion are less severe than in citizen-facing implementations. When one also considers the potential harm an employee can do when given access to resources without being held accountable for the misuse of these resources, the argument for employee authentication becomes compelling.

PC/Network Access

PC/network access is the use of biometrics to identify or verify the identity of individuals accessing PCs, PDAs, networks, applications, and other PC-oriented resources. The biometric is used to complement or replace authentication mechanisms such as passwords and tokens. More hardware and software solutions are available for PC/network access than for any other application. The widespread use of PCs and hand-held computing devices in home and work environments, the increasingly valuable resources available on corporate networks and the Internet, and the need for user-level authentication all contribute to make biometrics an obvious solution for PC/network access authentication.

Today's PC/Network Access Applications

Of the seven horizontal classifications, PC/network access is closest to traditional logical access. Logging onto PCs, laptops, and, eventually, hand-held devices is an obvious application for biometrics, a convenient and secure replacement for passwords. Accessing sensitive files, applications, networks, and databases, as well as other post-login resources, is another common application of biometrics in this arena.

Currently the only authentication spaces fully addressed by PC/network access are Windows NT/2000 and, to a lesser degree, Novell Netware. Biometrics are closely tied into Windows and Novell security systems and leverage the trust relationships and security infrastructure found in the operating system (OS). Dozens of hardware and software solutions that have been developed to this point offer authentication to Windows desktops and networks. These applications may be tightly integrated into the Windows security infrastructure or may reside as standalone authentication components logically removed from the Windows infrastructure. Biometric solutions deployed in these environments are generally client/server solutions, giving system administrators the ability to audit usage, manage security levels, and remove unauthorized users. Some solutions are deployed with client-only software, especially in mobile environments, to control access to the PC itself.

Solutions are emerging rapidly that allow deployers to integrate biometric functionality into Web sites, replacing password authentication. These solutions can be used for employee-facing intranets and, eventually, for customer-facing Internet applications. In addition, select biometric middleware offerings are capable of integrating biometrics into the username/password dialogue of various applications. These solutions are much less frequently deployed than Windows/Novell OS-level solutions. While a handful of biometric solutions have been ported to operate on Solaris, very few actual biometric deployments have taken place in this computing environment.

Both Microsoft and Intel have announced plans to incorporate biometric software in their respective authentication systems. This will be a critical driver of the technology's acceptance at the PC/network access level. As devices become more prevalent at the desktop, enabled by these major players, there will be increased demand for integration into email, customer relationship management, single sign-on, and applications specific to certain industries.

In many cases, integrating biometric solutions into enterprise applications requires custom development on the part of vendors, deployers, or firms dedicated to customized integration. While there are a number of off-the-shelf solutions, as the deployment environment grows beyond the standard Windows OS, it is more likely that custom integration will be necessary. Because of the proprietary nature of biometric technology, as well as the relative instability of many biometric vendors, some institutions have proven hesitant to fund the custom integration of biometrics into their operating environments. They prefer to wait until the technology has matured to the point where the return on their investment is more assured.

For home or individual users, there are a number of solutions on the market for PC security. Vendors who have targeted their products at the consumer market have not had a great deal of success because of the lack of awareness of biometrics as a solution and because of the cost of peripherals. Spending $100 for a peripheral device is seen as a reasonable investment for an enterprise but is slightly more than many home users are comfortable with.

From late 2000 into early 2001, a series of large-scale PC/network access implementations were announced, redefining *large-scale* in the PC/network access segment. Among these applications—all of which used finger-scan and constituted secured employee access to PCs—were the following:

- In January 2001, the city of Glendale, California, began a deployment to 2,100 city employees.
- In March 2001, the New York State Office of Mental Health announced that its PC/network access project, begun in 1998, had reached a staff level of 6,000.
- In early 2000, Credit Union Central, British Columbia, announced plans to expand its biometric rollout from 500 to 2,000 seats by the end of 2001.
- A U.K. production facility commissioned the deployment of a finger-scan peripheral for securing a 4,000-seat network.

Future Trends in PC/Network Access

Biometrics are expected to become a more prominent application in PC/network access due to the increased availability of data on shared networks, the increasing number of ways users expect to be able to access this data, and the sensitive nature of this data. Authentication at points where data is susceptible will become increasingly important. While this access currently relates primarily to desktop and mobile PCs, it is growing to include wireless, virtual private network (VPN), and any other application in which employees must access sensitive data.

The demand for stronger auditing and accountability for companies that store or manage sensitive data will be a large driver of biometric deployments in PC/network access. Beyond authentication, biometrics provide strong nonrepudiation, such that an audit trail indicating that a user was present at a given time can be used to minimize corporate liability.

The finalization of standards, including application programming interface (API), encryption, and file formats, removes an impediment to large-scale deployments. Potential deployers have been hesitant to commit to a technology that had not been through the process of formalized standards development. In 2000 and 2001, many of these standards, including BioAPI, BAPI, X9.84, and Common Biometric Exchange File Format (CBEFF), reached maturity and were being incorporated into product design.

The growth of smart cards as a PC-oriented solution for data storage and PC access could also drive revenues in this area, as biometrics and smart cards are highly synergistic technologies. Outside of the United States, smart cards have become a pervasive solution, though they have not yet become widely deployed for PC/network access. Because access to smart card data is often predicated on PIN usage, the integration of biometric functionality in the card access subsystem would allow for the storage of more sensitive data.

Along with smart cards, PKI is a technology that strengthens and is strengthened by biometrics for PC/network access. If and when certificate issuance begins to reach the user level and authentication to networks and resources occurs through rights and privileges embedded in certificates, the need for biometrics increases. PKI and certificate usage is strengthened and validated by the use of biometrics: PKI authenticates the machine; biometrics authenticate the user.

At the same time, a handful of impediments could slow the emergence of biometrics in PC/network access. Privacy and legal issues could make enterprise deployers hesitant to implement what can only operate as a mandatory system. Employees might protest based on the grounds of personal objections or even disabilities. Until some type of legal framework is established around the use of biometrics—not necessarily limiting its use, but defining what types of use require certain controls—deployers may be hesitant to embrace biometrics.

Major downturns in the PC market could delay the rate at which biometric devices are deployed to desktops (a precondition of robust PC/network access growth). Over the next one to two years, biometric devices will be incorporated in keyboards, mice, laptops, and so on; however, the replacement life cycle may grow so long that instead of a one- to two-year period elapsing as devices reach desktops, two to three years may be required.

In addition, while single sign-on and biometrics are not competing industries, widespread acceptance of single sign-on applications would remove one of the basic arguments for biometrics as a PC/network solution: the ability to eliminate password management. Biometrics may serve as a strong front end to single sign-on applications or may be negatively impacted if single sign-on becomes widely deployed.

PC/Network Access: Biometric Solution Matrix

Biometrics are a balanced solution to the problem of authentication in PC/network access, as indicated by the Biometric Solution Matrix. The main area limiting the revenue potential of biometrics in this application is exclusivity, as biometrics will face various competitors in this segment (see Figure 12.1).

Exclusivity (6/10). There are a number of technologies capable of providing authentication at the PC/network access level: tokens, cards, and, most important, passwords. Tokens and cards, while arguably more reliable than biometrics, have the disadvantage of being shareable and capable of being lost. While biometric vendors attempt to position the password as a major problem for enterprises, the truth is that the password is a reasonably effective authentication method used by hundreds of millions of individuals every day. The issues involved in password usage are known quantities, whereas the use of biometrics raises questions that are not yet fully answered, such as performance over time and technology obsolescence. Biometrics are not the only solution in PC/network access, just the most robust.

Effectiveness (8/10). Biometrics are an effective solution for PC/network access, providing fairly low error rates and fitting well within existing authentication processes. The PC/network access environment is fairly forgiving, with consistent environmental conditions, a usable environment well suited to the introduction of a peripheral, and motivated users. The biometric technology most frequently used in PC/network access, finger-scan technology, has been optimized for performance in this environment and is, by biometric standards, mature and proven.

Receptiveness (8/10). Because PC/network access biometric solutions are often deployed to replace passwords while providing increased security for sensitive data, there is a general receptiveness to the technology on the part of both deployers and those being authenticated (primarily employees). A receptive user group is by no means guaranteed; if the technology is deployed without explanation of the system objectives and a description of how biometric systems operate, users are more likely to react negatively.

Continues

PC/Network Access: Biometric Solution Matrix (continued)

Urgency (7/10). Although PCs and networks have operated effectively for years without biometrics, there is a heightened awareness of the risks of unauthorized access to data. Recent incidents in which laptops with classified information have been lost and their data presumably compromised have brought attention to the susceptibility of data residing on PCs and networks. While the highest-visibility hacking incidents involving loss of credit card and personal data on Web servers would likely not have been prevented by biometrics, the problem of network and PC security is viewed as increasingly urgent. Furthermore, the risk of data loss due to password compromise in this environment will only grow more acute.

Scope (8/10). Biometric solutions for PC/network access have the ability to affect the day-to-day lives of individuals at work and at home. Nearly every PC and, over time, handheld device will be a suitable deployment environment for a biometric solution. Breadth of scope is a key factor in moving biometrics from being a niche solution to being a wide-ranging solution.

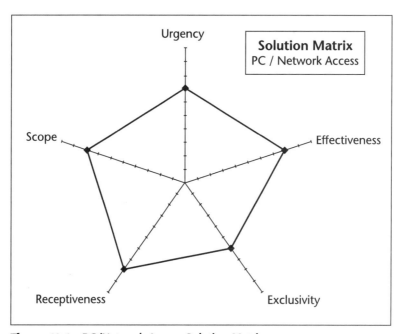

Figure 12.1 PC/Network Access Solution Matrix.

Related Biometric Technologies and Vertical Markets

Finger-scan and middleware are the two technologies most strongly associated with PC/network access. Many finger-scan devices provide the high levels of accuracy, small form factors, and ease of use necessary for PC/network access deployment. These devices have become increasingly suited to desktop applications over the past several years: At first cumbersome and expensive, today's finger-scan devices optimized for PC/network access are built into keyboards, mice, PCMCIA cards, and very small peripherals. Finger-scan solutions are generally deployed with software developed by device manufacturers or with middleware designed specifically for enterprise security.

A number of other biometric technologies have attempted to penetrate this segment, including facial-scan and voice-scan. Adoption of these technologies has been nominal, as they are not well suited to performing in desktop environments. Background noise, background lighting, and changes from desktop to desktop make these technologies susceptible to high false rejection rates. Iris-scan technology is being positioned as a solution for the PC/network access environment, potentially being able to overcome some of the performance issues that affect other technologies, but it has not yet been deployed widely.

There are a number of verticals for which biometric PC/network access is a critical solution, including healthcare, financial services, and government-sector applications. However, biometric technology is a robust PC/network access solution in any industry—the only preconditions are desktops with sensitive information. PC/network access is likely to be one of the strongest growth areas of the biometric industry, because it provides authentication to the operating systems and software that reside on nearly all enterprise and home desktops.

Costs to Deploy Biometrics in PC/Network Access

The cost to deploy biometrics as a PC/network access solution, especially in an enterprise environment, varies according to the number of users, the complexity of the authentication environment, and the type of biometric solution deployed, including hardware and software. Most enterprise deployments involve the purchase and installation of peripheral devices, whose costs can be borne up front or paid over time on a contractual basis. Fortunately, hardware costs have fallen sharply in recent times—there are a number of high-quality $100 finger-scan devices, and these prices are expected to drop as biometrics become integrated into keyboards and other peripherals.

In addition to the per-desktop hardware costs, software may have to be licensed separately, either per user or per seat. While peripherals often ship with local login software, most enterprises require centralized control over user authentication, meaning that additional software costs are involved. For enterprise deployments, authentication software packages tied to a specific hardware solution ranging in cost from $500 to $1,500 enable central storage, verification, and management of biometric data. Additional per-seat fees are involved once a certain number of users are enrolled. Middleware solutions that enable a variety of biometric and nonbiometric authentication devices are priced on a central server and seat license basis. Middleware server pricing and per-seat prices can be significantly higher than those for standard device-specific software because institutions are investing in an infrastructure capable of using various authentication methods. For middleware deployments, there may also be additional costs for customized integration or implementations.

A 1,000-seat finger-scan deployment for Windows 2000 login, for example, might cost an enterprise between $100,000 and $250,000, depending on the type of device deployed and whether the deployment utilizes vendor-specific authentication software or more expensive and flexible middleware. This figure does not account for the time required to research, evaluate, and test solutions; implement and administer the system; and train and enroll users. Though these costs are not terribly high when one considers the value of assets being protected, it is easy to understand why voice-scan and keystroke-scan, because they are capable of leveraging existing acquisition devices, may eventually be appealing lower-cost solutions.

Companies are looking to bring service and subscription models to enable PC/network solutions, an effort to generate a consistent revenue stream while lowering initial deployment costs for enterprises. In these situations, the biometric verification might be outsourced via the Internet or hosted internally by the deployer. This market has not stabilized enough for prices to have come into focus, but a monthly fee in the area of $8 to $15 per user per month for outsourced authentication has been proposed.

Issues Involved in Deployment

Aside from the basic considerations a deployer must consider when determining whether and how to use biometrics, there are a number of specific considerations to bear in mind for PC/network access.

How and when will enrollment take place? Initial user enrollment in a large-scale PC/network access system can pose severe logistical challenges. While many PC/network solutions are capable of enrolling users at their desktops, avoiding the logistical challenge of scheduling users at a central enrollment location or at several enrollment hubs, some require enrollment at central administrative stations. Deployers must ensure that the enrollment process at these distributed locations mirrors the user's desktop environment, or false rejection rates may be higher than anticipated. Regardless of where enrollment takes place, deployers must ensure that only authorized users are enrolling through some type of password or identity check, or else the initial integrity of the system can be compromised. Also, the quality of enrollment is essential to ongoing verification in the system. Marginal quality enrollments are more prone to falsely rejecting authorized users and falsely accepting imposters.

Do users remain at one desktop or roam from PC to PC? If users roam from PC to PC, a centralized authentication solution must be deployed that is capable of allowing authentication from any desktop. In addition, the suitable hardware must be present at any desktop at which the user might need to verify. Without the device, users will be forced to authenticate through password, and the system becomes susceptible to attacks. Along the same lines, if PCs are shared, the solution deployed must be capable of authenticating multiple users to the same PC.

Do users access resources from remote locations? Increasing numbers of employees access protected resources from a variety of locations—work, home, and travel. Authentication solutions must be similarly capable of operating in these conditions, meaning that compatibility with dial-up service and VPNs may be a necessity. Because of the fairly limited biometric solutions compatible with dial-up and VPN operation, the solution deployed for mobile users is often to use the biometric at the complementary metal oxide semiconductor (CMOS) or boot-up level. These solutions require a successful authentication before the operating system even loads, ensuring that protected resources cannot be accessed.

What fallback procedures are in place? Biometric systems, on occasion, reject authorized users; these systems also reject imposters. Unfortunately, the technology cannot distinguish between the two. The policies established by which rejected users are processed are essential to secure and convenient system operations and are especially important in a PC/network access environment because of the mandatory nature of these implementations. A common first procedure is to enroll more than one biometric sample, if possible (the ability to enroll multiple samples is a strength of finger-scan technology). In this case a user can fall back to a secondary finger for verification. In many cases, the solution of false rejection may be to fall back to a password. Of course, imposters can then intentionally be rejected by a system and gain access by attacking the password. Administrators might monitor password-authenticated users more closely, enforce more complicated policies, or reset passwords after a single use. One of the difficulties in developing fallback procedures is that fallbacks cannot be punitive. Some users are more likely to be falsely rejected due to age, ethnicity, or occupation, such that imposing onerous fallback procedures could be seen as discriminatory.

How will the system be introduced to users? Since a cooperative and informed user population is essential to successful deployment in a PC/network access environment, users must be familiarized with the system and its basic operations. Because users may view the technology as being an imposition or invasion of privacy before they understand its rationale and operations, it is essential to conduct some sort of orientation that explains basic concepts such as matching, templates, and system errors. It is equally important to explain why the system is being deployed, whether for increased security or decreased administrative costs.

How will security levels be established? Centrally administered PC/network access systems normally have flexible security levels. This allows deployers to establish security as their primary concern, which limits the likelihood of false acceptance but increases the likelihood of false rejection, or to establish convenience as the primary driver, which limits the likelihood of false rejection but increases the likelihood of false acceptance. In lieu of actual performance data, trial and error is often the only way to determine the meaning and accuracy of vendor-supplied security settings.

Conclusion

PC/network access is likely to be one of the first areas in which individuals interact with biometrics on a daily basis. Though an emerging biometric market, the solutions available for PC/network access are more mature than those in many other biometric applications, meaning that the risks of deployment are lower than with other biometric technologies. As Microsoft, Intel, and other large corporations build biometric logic into their authentication infrastructures, usage should continue to grow less expensive and more simplified.

Physical Access/Time and Attendance

Physical access/time and attendance is the use of biometrics to identify or verify the identity of individuals entering or leaving an area, typically a building or room, at a given time. The biometric is used to complement or replace authentication mechanisms such as keys, tokens, and badges.

Today's Physical Access/Time and Attendance Applications

The first commercialized uses of biometric technologies were in physical access, as devices were deployed to control access to secure areas of military facilities, power plants, and banks. (see Figure 12.2.) The logic behind biometric authentication is highly compelling in this environment, as keys and badges are easily shared without being traceable to the actual user, while biometrics cannot be shared without the complicity of an enrolled user. The use of a biometric creates an audit trail that is difficult to repudiate.

Physical access solutions are normally deployed to select rooms in a facility, building, or office environment, and are rarely used to control access to every door.

Figure 12.2 Physical access devices.

Time and attendance is also an area in which strong authentication is a necessity, although the rationale is not to control access to sensitive areas but to detect and deter fraud. Successful deployments of biometrics in physical access environments have saved companies hundreds of thousands of dollars in reduced buddy-punching—clocking in a coworker without him or her being present. Time and attendance solutions are often, but not exclusively, tied into physical access systems. Time and attendance systems can be installed in break rooms, at building entry, and so on, without controlling access to secure areas.

Future Physical Access/Time and Attendance Trends

Growth in the physical access/time and attendance market will be driven by faster, cheaper, more accurate, 1:*N*-enabled devices. Without identification capabilities, one of the primary benefits of biometrics—not having to remember a token or PIN—is lost. Biometrics have not reached truly widespread acceptance in physical access environments because, in many cases, their cost and relative ease of use do not provide a compelling argument to replace existing systems. Reduced cost and increased ease of use will drive growth in select applications.

The time and attendance market seems to have slightly more upward potential and may be able to drive physical access growth if the two are packaged together. The sizeable reduction in fraud attributable to biometric time and attendance devices detecting and deterring fraud has been demonstrated in numerous large-scale applications. If biometric vendors can integrate these capabilities with other human resource applications, there may be prospects for more revenue generation than currently attained in this segment.

While device-level improvements should drive more employers to embrace physical access/time and attendance solutions, revenue growth may be limited by the "multiple users, one device" relationship. As opposed to other horizontals such as PC/network access, in which each user needs a device on his or her PC, a handful of physical access devices can accommodate a large number of users, limiting revenue potential. In addition, physical access/time and attendance solutions cannot be readily expanded to embrace new functionality.

Furthermore, the strength of competing technologies in this horizontal represents a significant growth inhibitor. Companies concerned with security have several solutions to choose from; integrators and installers often steer would-be deployers away from biometric solutions to simpler solutions such as proximity cards.

Related Biometric Technologies and Vertical Markets

Hand-scan and finger-scan are the most commonly deployed physical access/time and attendance solutions. Hand-scan has been in use since the 1980s and is deployed in tens of thousands of locations, while finger-scan's higher accuracy and greater flexibility have given it traction in this market. For the highest-security applications, iris-scan and retina-scan technology have been deployed, due to their very low false match rates. However, retina-scan became a much less prominent biometric technology in 2000 and 2001 production, while iris-scan was reoriented as a desktop solution and de-emphasized as a physical access solution.

In terms of vertical markets, biometrics are deployed in a range of physical access environments from low to high security, including health clubs, schools, casinos, and power plants. There are no specific environments for which biometric physical access solutions are better equipped, although the transportation industry has deployed a number of biometric solutions for employee access control.

Costs to Deploy Biometrics in Physical Access/Time and Attendance

The costs of implementing physical access solutions are found primarily in hardware and integration, and to a lesser degree in software. Primarily implemented as door-control devices and tied into existing building management systems, physical access/time and attendance solutions range in price from as low as $600 for some finger-scan devices, to approximately $1,500 to $2,000 for hand-scan devices, to upward of several thousand for iris-scan systems. These devices are normally sold in small quantities through distributors and are professionally installed to ensure proper communication with existing security systems. Integration costs are dependent on the number of devices being sold, the complexity of the existing security system, and whether a great deal of modification needs to be done to doors and walls.

For deployments in which several points of entry are controlled through a single management system, PC-based software is available that allows administrators to collect and distribute between devices, eliminating the need to enroll at multiple locations. This software is fairly inexpensive, often less than $1,000.

Time and attendance solutions are generally more software oriented, with the biometric devices being tied to a central logging or auditing system that translates the authentication data to corporate payroll systems. The costs of these

Physical Access/Time and Attendance: Biometric Solution Matrix

As shown in the Solution Matrix, biometric technology is a moderately capable solution for physical access/time and attendance, but there are some issues to be overcome that may limit its advancement. These limiting factors include exclusivity, effectiveness, and urgency (see Figure 12.3).

Exclusivity (5/10). Biometrics are *not* a highly exclusive solution in physical access/time and attendance: The technology competes with badges, tokens, cards, even standard keys carried by many employees. Although these physical devices can be shared, reducing overall system security, they are generally easier to use than biometrics in many physical access environments. It will be a challenge for biometrics to uproot these less expensive challengers within a physical access solution. In time and attendance, biometrics are a slightly more exclusive solution, as biometrics are the only solution capable of fully solving the buddy-punching and time card fraud problem.

Effectiveness (6/10). Biometrics work reasonably well in physical access/time and attendance environments, but the lack of rapid-response identification-enabled units has hampered its growth. Compared to proximity cards, which allow entry with almost no delay, most biometrics require the presentation of a card or PIN, followed by careful presentation of a biometric. Without identification systems for physical access, the solution's effectiveness is not ideal. The efficiency of biometrics in time and attendance applications is higher, as ingress is not as pressing an issue.

Receptiveness (7/10). Physical access/time and attendance solutions are almost invariably aimed at employees, who are moderately receptive to the introduction of the technology. Generally speaking, physical access solutions will be viewed more kindly than time and attendance, as time and attendance solutions increase personal accountability, implying that some employees require this stronger type of authentication. Receptiveness will decrease if solutions do not operate as easily and reliably as the existing authentication method.

Urgency (6/10). A compelling need for robust physical access solutions is present in specific verticals, such as air travel and schools. Institutions are looking to ensure that they have taken all possible precautions to protect their resources and to control sensitive areas—an issue that has become more visible

Continues

Physical Access/Time and Attendance: Biometric Solution Matrix (continued)

since the September 2001 attacks on the United States and the increase in workplace and school violence. Time and attendance solutions are also viewed as an urgent need in specific verticals, such as retail, where employee fraud can significantly impact profit margins.

Scope (8/10). The strongest positive factor for physical access/time and attendance growth is the very broad range of potential uses of the technology. The use of biometrics to control movement into areas or to track attendance is not a niche market—there are potentially countless applications for biometric technology. Not only are the potential locations for these solutions innumerable, but the number of employees who would be enrolled in such systems is also very large.

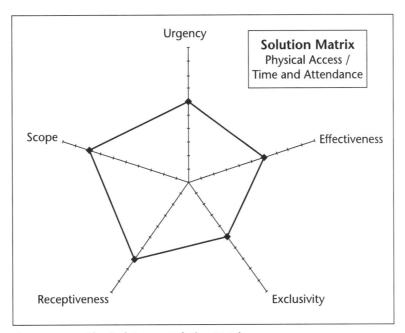

Figure 12.3 Physical Access Solution Matrix.

systems are based on hardware, software, and whatever custom integration may be necessary to integrate both the physical and logical components with existing systems.

Issues Involved in Deployment

There are specific challenges involved in determining how to deploy biometrics effectively in a physical access/time and attendance environment.

What is the current access control system? Depending on the current access control system, biometrics may be just slightly more cumbersome or much more cumbersome. Users accustomed to waving a proximity card at a reader, barely breaking stride, will find that the biometric system requires much more effort and concentration, especially when deployed in verification mode as opposed to identification mode. This may result in an increased willingness for users to hold open secure doors, as the biometric system is viewed as an annoyance. For time and attendance solutions deployed independently of physical access solutions, this is less of a concern, as users have no option but to authenticate biometrically.

What is ingress/egress like at the locations being secured? The problem of ingress and egress at access points can hinder the feasibility of deploying biometrics. If a number of users pass through a biometrically controlled access point at a given time, such as at the beginning of a shift, the biometric must authenticate users quickly enough to allow the next user to authenticate quickly. This authentication process includes the time necessary to enter any PIN numbers for the initial identity claim or to swipe a card, as well as the time necessary to provide biometric data and receive a response. If the process is too time consuming, it is possible that users will circumvent the physical access point by some means.

What fallback procedures are in place? As with PC/network access solutions, there will be some percentage of physical access users who are unable to verify successfully on a daily basis, attributable to failure to enroll or to false rejection. Whereas fallback procedures of varying strength can be easily employed at the desktop, it is more difficult to implement fallback procedures for physical access. While users unable to enroll might be assigned PIN codes that they can use to verify—assuming that the devices in place utilize PINs—users unable to verify are a greater challenge. Factors such as weather and environment can lead to increased error rates for deployments exposed to the elements. In these situations, there will generally not be recourse to a help desk, given the nature of physical access usage, and there are few readers capable of processing both access control badges and biometric authentication. Users might be assigned a one-time code to use if they cannot verify on any enrolled finger, for example, and then reenroll at a different enrollment threshold or security level.

Conclusion

Physical access is a strong biometric application for specific environments, offering substantial improvements over many existing solutions. It is less likely to enjoy pervasive deployment than PC/network access solutions, because certain doors and access points are too busy to sustain the few-second authentication process required by verification devices. As identification solutions become more prevalent, it is likely that the number and scope of physical access deployments will increase. Time and attendance, in comparison, should be a more rapidly growing solution, because there are fewer existing processes with which the deployment of biometrics can interfere.

The core technology may be identical when moving from employee-facing deployments to customer-facing deployments; however, Chapter 13 demonstrates how the challenges of customer-facing applications differ significantly.

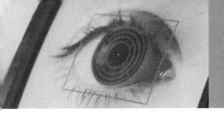

CHAPTER 13

Customer-Facing Applications

Customer-facing applications include e-commerce/telephony and retail/ATM/point of sale (POS). The defining element of customer-facing applications is that a provider of goods or services provides authentication of consumers and enforces compliance with the biometric system's match decisions. Customer-facing applications are generally open systems, incorporating some percentage of a seller's customer base.

Customer-facing applications are almost always optional and may or may not be predicated on centralized verification. Customer-facing applications are normally based on verifying a claimed identity, not on identification. When biometrics are implemented for identification in this environment it is for convenience—allowing users to authenticate without having to recall a user name—as opposed to security. In contrast to employee-facing applications, there is no degree of coercion involved in customer-facing deployments; if anything, customers might be rewarded for enrolling. The privacy and system design implications of customer-facing implementations change accordingly. While potentially the most lucrative field for biometric deployments, customer-facing implementations are the least developed of the three capacity-related classifications.

E-Commerce/Telephony

E-commerce/telephony is the use of biometrics to verify the identity of individuals conducting remote transactions for goods or services. The biometric is used to complement or replace authentication mechanisms such as passwords, PINs, and challenge-and-response interaction. Though there are differences between e-commerce and telephony—primarily in the acquisition technologies that enable each application—e-commerce/telephony is a logical grouping for several reasons.

Both applications involve remote parties and typically lack a high degree of trust. There is a need to establish the identity of the agents in both e-commerce and telephony applications; without biometrics, it is difficult to do so efficiently and reliably. Similarly, both currently lack secure enrollment processes, such that lower degrees of certainty might be associated with certain enrolled users.

Both applications are also largely oriented toward transactional usage, meaning that the two applications may develop similar pricing and revenue models. The primary difference—that one leverages a pervasive acquisition infrastructure (telephones) and the other requires that an infrastructure be deployed (e.g., finger-scan readers)—will diminish as finger-scan devices are more commonly deployed. Finally, both applications lend themselves to the introduction of a trusted third party capable of processing transactions and routing responses to the appropriate remote locations.

Today's E-Commerce/Telephony Applications

To this point, there has been no typical e-commerce application, as there are few commercial Web sites that utilize biometric authentication for customers or clients. The few existing implementations are in telephony, where a small number of financial institutions secure account access through biometric verification. Over the next one to two years, a typical application will emerge consisting of transactional biometric authentication—that is, an individual placing a finger or speaking a passphrase in order to execute a transaction. These applications will occur, at first, in B2B environments, with the intent of creating a strong audit trail for high-value transactions. Over time, typical e-commerce/telephony applications will secure a broader range of transactions.

Perhaps the most distinguishing characteristic of this market is that e-commerce and telephony solutions will not be *deployed* so much as *utilized*. To illustrate, employers deploy hardware and software to their employees and deploy infrastructures through which these devices can operate. Aside from proof-of-concept pilots, merchants do not deploy devices to their customers, but instead will utilize devices already in use through authentication infrastructures.

Future E-Commerce/Telephony Trends

Though the initial euphoria over e-commerce has faded, there remain several compelling reasons why e-commerce/telephony should be a major growth area for biometric technology.

Individual and Institutional Need for Increased Trust in Remote Transactions

E-commerce/telephony is a fairly unique application in that both agents—the seller and the purchaser, in most cases—are positively motivated to utilize and require strong authentication. As opposed to network security, for example, where institutions are more strongly motivated to implement biometrics than employees are to use them, both parties to an e-commerce or telephony transaction derive benefit from the introduction of a biometric. The seller, or information holder, benefits by decreasing the risk of being defrauded; the buyer can be certain that only he or she can authorize transactions on the protected account.

Emergence of Transactional Revenue Models

Revenues in the biometric industry have been driven by hardware sales, software licensing, and service and support. There are very few transactional revenue models in the industry, outside of offerings from select voice vendors and large-scale AFIS systems. The advent of a transactional revenue model—enabled by a transactional infrastructure that mirrors the existing transactional infrastructure in e-commerce—will likely drive substantial growth in the biometric industry. As a whole, the biometrics industry currently views its solutions as devices that generate one-time revenue or as infrastructures that generate subscription-based revenues. While these revenues will continue, more interesting revenue streams will emerge when devices are used several times per day to various applications, authorizing transactions, sales, and purchases. Instead of a one-time seat and device sale for $150, the device—when enabled by a transactional infrastructure—becomes a conduit for ongoing revenue generation.

Increased Presence of Acquisition Devices

A precondition of e-commerce/telephony revenues is the broad availability of acquisition devices. For telephony, acquisition devices are already present everywhere and will only become more prevalent. For e-commerce, finger-scan devices (expected to be a common solution for desktop applications) will initially be deployed as PC/network access solutions and then see increased utilization as e-commerce solutions.

It should be emphasized that e-commerce and PC/network access, while different applications of biometric technology, are each based on the utilization of devices such as finger-scan readers on user desktops and laptops. The differences are that (1) in PC/network access, the relationship between the authenticator and the authenticated is generally employer to employee, while in e-commerce, the parties are buyer and seller, and (2) revenues in PC/network access are based on initial system sales, while revenues in e-commerce are transactional.

Entry of Trusted Parties into the Biometric Market

E-commerce/telephony will also be characterized by the entry of trusted third parties into the biometric space. There are no vendors of just biometrics that are large or established enough to qualify as trusted; large companies that manufacture biometric devices have not traditionally acted as trusted third parties. Companies with a reputation for processing transactions, storing data, and acting as transaction intermediaries will enter this market space, providing levels of trust that are currently lacking between unknown parties in the biometric industry.

There are also potential restrictions on the emergence of biometrics as an e-commerce/telephony solution. First, biometrics must reach the desktop, meaning that major players such as Microsoft and Intel must continue down the path of integrating biometric functionality into their core software offerings. Without the infrastructure to enable e-commerce transactions, growth in this market cannot occur.

If trusted third parties fail to emerge or there are problems with the ability of third parties to provide broad compatibility with devices resident on user desktops, revenues will be negatively impacted in this space. Specifically in telephony applications, if initial deployments for transactional authentication reject authorized users or accept unauthorized users, the negative reaction from customers may delay other potential deployers (see Figure 13.1).

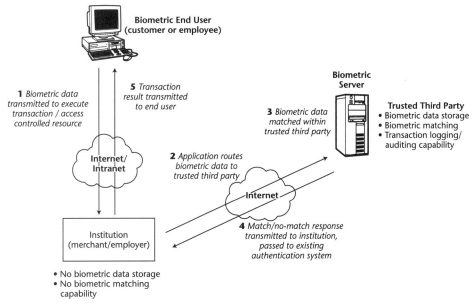

Figure 13.1 Role of trusted third parties.

Related Biometric Technologies and Vertical Markets

Finger-scan is likely to be the most frequently utilized technology in e-commerce applications, assuming that the technology reaches the desktop as anticipated. Other technologies, including iris-scan and facial-scan, may also be utilized, but finger-scan has a substantial head start in this area. In the telephony portion, voice-scan will dominate, though there is a possibility that mobile telephony devices with finger-scan devices will gain some degree of acceptance.

Financial services and healthcare are two of the obvious vertical markets in which e-commerce and telephony are necessary solutions, as account access, financial transactions, and authorization of medical services are all areas in which remote authentication is a necessity.

Costs to Utilize Biometrics in E-Commerce/Telephony

Biometrics have yet to be utilized on a large scale in e-commerce implementation and have only found limited deployment in telephony. Because of this, the costs for merchants to utilize biometric authentication in customer-facing applications have not been firmly established. Whereas employers' costs are based on the number of users, the type of device deployed, and the complexity of the authentication environment, merchants face a different set of variables: With which devices will they be compliant? What infrastructure will they utilize to enable biometric solutions? How will the fees and liability of biometrically authenticated transactions differ from standard transactions?

It is likely that merchants will pay flat or percentage-based fees for the benefit of having biometrically authenticated users, a group much less likely to commit many types of fraud than nonbiometric users. A portion of these fees will most likely go to core biometric vendors, whose technology will be licensed to perform biometric matching. However, the greater portion will go to transaction enablers who can provide authentication for merchants who would otherwise need to develop and maintain these biometric competencies internally.

A very plausible cost scenario involves merchants paying transactional, user-based, or percentage-based fees to biometric authentication providers. Just as financial institutions and clearinghouses exist to assume the burden of purchaser authentication from merchants in credit card transactions, biometric authentication providers will allow merchants to process biometric transactions to which they would otherwise not have access. These authentication providers may provide internal software solutions for merchants, or they may provide outsourced functionality.

In the more established telephony component of the market, merchants and institutions can either implement internal solutions, often priced in the tens of thousands of dollars, or implement outsourced solutions, which route callers to third parties for authentication. Telephony infrastructures do not need to be compatible with dozens of biometric devices, as would an e-commerce infrastructure. A single voice vendor can enroll and authenticate a broad range of users because the impediment of multiple acquisition devices found in e-commerce does not exist in telephony.

The cost of implementing customer-facing e-commerce and telephony infrastructures may also vary according to fees associated with high-value, high-risk, or high-security transactions. Standard authentication technologies cannot provide the granularity that biometrics can provide in terms of variable security levels. High-value transactions can require higher degrees of correlation in biometric data, and fees can be imposed accordingly. In many cases, the presence of a biometric will enable higher-risk and higher-value transactions than are generally executed in remote environments.

E-Commerce/Telephony: Biometric Solution Matrix

The E-Commerce/Telephony Solution Matrix demonstrates why biometrics are a strong solution for this sector. With the exception of exclusivity, biometrics rank highly in each of the categories that define the suitability of biometrics for a given application. (see Figure 13.2.)

Exclusivity (6/10). Biometrics are not a highly exclusive solution in e-commerce and telephony, competing with numerous authentication methods of varying robustness. PINs, passwords, and challenge-and-response mechanisms are commonly used to authenticate individuals in this environment. Transactions are currently executed in both e-commerce and telephony using these traditional authentication mechanisms; what is lacking is a high degree of trust.

Effectiveness (7/10). Biometric solutions are capable of performing well in most e-commerce/telephony environments. In telephony, voice-scan is an effective solution when enrollment and verification each take place on a high-quality line, with many systems providing high levels of accuracy without interrupting existing process flows. Finger-scan is viewed as the most suitable solution for e-commerce because of the variety of desktop form factors and its high degree of accuracy. Both e-commerce and telephony environments are well suited to the introduction of biometric technology.

Receptiveness (9/10). Though the deployment of technology in the e-commerce/telephony environment will be optional, not mandated, both parties to these transactions are motivated to utilize strong authentication technologies. There is a high degree of receptiveness to the deployment of biometrics in this application, such that one of the impediments present in most biometric systems—having to convince an individual or institution of the value of implementation—is not present.

Urgency (8/10). There is a moderately high, but not overwhelming, need for a biometric solution to the problem of authentication in e-commerce/telephony applications. Solutions exist that address e-commerce and telephony applications, but many are unreliable or limit the value or levels of trust associated with transactions. This problem will grow more urgent over time as more valuable remote transactions require authentication.

Scope (9/10). The potential scope of biometrics as a solution for e-commerce/telephony authentication is very broad—not only are there millions of potential transactors, but each person might have multiple types of transactions in which he or she can authenticate. One of the reasons that e-commerce and telephony are such appealing markets for biometric companies is the potential scope of these applications.

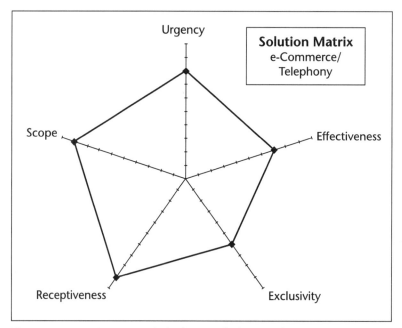

Figure 13.2 E-Commerce/Telephony Solution Matrix.

Size of the Biometric Industry

The biometric industry is small but rapidly growing. Measured in terms of revenues based on the sale of biometric hardware, software, and vendor-provided customization, the total size of the industry in 2001 was over $500 million. However, more than 60 percent of this revenue was attributable to AFIS revenues. Non-AFIS revenues, including revenues derived from hardware sales and software sales, comprised slightly over $200 million (see Figure 13.3).

MARKET SHARE BY TECHNOLOGY

Estimates for 2001 show that finger-scan continues to be the leading biometric technology in terms of market share, commanding nearly 50 percent of non-AFIS biometric revenue. Facial-scan, with 15.4 percent of the non-AFIS market, surpasses hand-scan, which had for several years been second to finger-scan in terms of revenue generation (see Figure 13.4).

Customer-Facing Applications 197

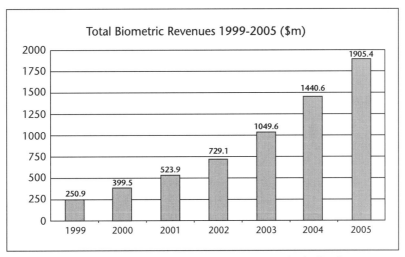

Figure 13.3 Total biometric revenues 1999 to 2005 (including law enforcement/AFIS usage).

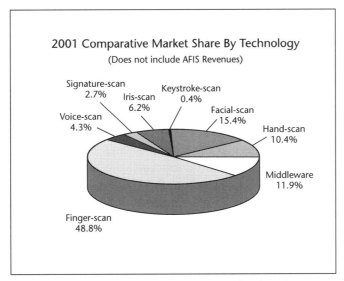

Figure 13.4 Biometric market share by technology for 2001 (excluding AFIS).

Issues Involved in Deployment

The following are among the key issues institutions must consider when implementing biometric systems for e-commerce and telephony.

Enrollment

One of the defining characteristics of e-commerce/telephony applications is that nearly all users will be enrolled remotely. Since during remote enrollment individuals are unsupervised, there is an increased risk of both fraudulent enrollment attempts and poor-quality legitimate enrollments. A nonbiometric authentication process, preferably more robust than simply name, date of birth, and social security number, is required to ensure that the individual enrolling biometric data is who he or she claims to be. Solutions might include mailing a one-time password or requiring submission of thorough background check data. Fraudulent enrollment clearly undermines the entire system, as a nonauthorized enrollee can effectively hijack a user's account. Low-quality enrollment undermines the system in another way. When users enroll in supervised environments, controls can be enabled that ensure that users are submitting data properly and that the biometric enrollment meets a minimum quality score. This decreases the likelihood of false rejection. With low-quality enrollments, not only are false rejections more likely, but in some systems false acceptances may also increase, as the enrollment templates are less distinctive. Most systems are designed to verify users immediately after enrollment, such that very low-quality enrollment templates can be avoided.

Another enrollment-related question is that of sharing enrollment data among institutions, most likely through trusted third parties. Requiring that a user enroll separately for each of several e-commerce or telephone applications is counterintuitive—once a user is enrolled the first time, it seems more logical that this enrollment should serve as the basis of future authentication across a range of systems. Transactional authentication providers could play the trusted-third-party role, allowing enrollments in certain systems to be used as the basis of verification in other systems. This would require the consent of the merchant and, more important, the user, who would see the benefit of leveraging an existing enrollment but may be suspicious of potential data misuse.

Liability

Institutions implementing e-commerce and telephony solutions must address the problem of liability for fraudulent account access. Regardless of system security, the odds are overwhelming that at some point even robust biometric systems will falsely match an unauthorized user, resulting in the fraudulent approval of a transaction. In this case, it must be determined whether financial restitution is the responsibility of the merchant or the biometric authentication provider. If the e-commerce and telephony solution deployed is outsourced, such that the actual authentication is not under the direct control of the institution, it is more likely that the biometric authentication provider may bear some liability.

Fallback Procedures and Locking Accounts

In e-commerce/telephony implementations, the biometric system will reject the vast majority of imposters but will also reject a small percentage of authorized users. To avoid inconveniencing legitimate users, fallback procedures must be implemented by which nonbiometric authentication is possible. For telephony implementations, this would normally entail the call being routed to an operator, who can ask background questions in order to authenticate the customer. The problem is more difficult in e-commerce, as individuals executing transactions from home or work must be authenticated in a different fashion. For finger-scan users, there is always the option of using a backup fingerprint. Password fallback is another possibility, although all of the weaknesses of password authentication are then brought to bear. A challenge-and-response mechanism might be put in place, after which a reenrollment takes place to address future verification problems. This is complicated by the fact that users have come to expect rapid authentication for online transactions.

After a series of unsuccessful authentications, merchants may require that an account be locked and that no more biometric verification attempts be allowed. This reduces the likelihood that a specific account can be compromised.

Compatibility with New Devices and Technologies

It is very likely that a variety of biometric technologies and devices will be deployed at the desktop level. These devices have traditionally been incompatible, such that an enrollment on one device cannot be verified on another device. Institutions interested in authenticating their customers through biometrics must determine how to ensure compatibility with the widest range of devices, while at the same time ensuring that devices meet certain minimum standards of accuracy and performance. Institutions will also need to ensure that the infrastructure they utilize to reach customers will be compatible with new devices as they reach the desktop.

Integration into Existing Systems

Though biometric authentication will become more prevalent at the desktop and through telephones, e-commerce and telephony systems will retain their current authentication processes and routines. The biometric authentication will leverage, not replace, these back-end systems. To illustrate, envision a time when 80 percent of customers are using passwords, and 20 percent are using biometrics. While biometric transactions may be flagged to generate special pricing or processing instructions, at the most basic level the match provided by a biometric will still need to be indistinguishable from the match provided by a password. Having to reengineer either e-commerce or telephone systems to accept a variety of match types will likely be very cumbersome, as institutions may have a series of legacy systems developed over time. The biometric must be capable of feeding its responses into the current authentication logic.

Security Levels

Customer-facing solutions for e-commerce/telephony must balance the need to prevent imposter attacks with the need to match valid users. Institutions must determine the degree of risk they are willing to assume in either direction, as adjusting security levels will lead to increased likelihood of either false matching or false nonmatching. If the system is well designed, utilizing secure fallback procedures, it should be capable of minimizing false rejections while not becoming overly inviting to imposter attacks. One of the challenges institutions face is that there are very few independent measures of accuracy in the biometric industry—vendor claims are often based on laboratory testing of algorithms as opposed to testing of actual users.

Internal or External Infrastructure

As the e-commerce/telephony markets emerge, institutions will have the option of maintaining an internal infrastructure capable of providing customer-facing authentication or outsourcing this authentication to an outside company. There are advantages and disadvantages to both approaches. Maintaining an internal infrastructure reduces reliance on third-party operations and ensures that customer data is handled in a fashion consistent with merchant policies. Performance issues can also be addressed more readily, as the PCs providing biometric authentication are under institutional control. On the other hand, internal infrastructures require that an institution build an ongoing competence in storing and processing biometric data and that institutions be knowledgeable about device-level accuracy and performance. Institutions also need to go through the process of enrolling their own users and handling problem calls.

External authentication infrastructures allow institutions to focus on their core capabilities and to adopt biometric authentication with minimum effort, eliminating the need to acquire and manage biometric data and to stay abreast of new developments in the biometric market. Costs are generally more predictable, as external authentication is normally based on recurring fees. Reliance on external authentication also entails reliance on a third party's ability to manage and secure personal data in a responsible fashion; institutions are often hesitant to surrender customer information to third parties. Also, the institution is at the mercy of the third party's ability to process transactions rapidly, to provide reliable device-level accuracy and performance, and to remain a reliable partner.

Conclusion

Perhaps the most speculative and yet potentially lucrative biometric application, e-commerce/telephony is characterized by the use of biometrics to solve problems inherent in remote authentication between nontrusted parties. The need for increasingly high-value transactions and data exchange between nontrusted parties should position biometrics as a strong solution for merchants and institutions interacting with customers.

Retail/ATM/Point of Sale

Retail/ATM/point of sale (POS) is the use of biometrics to identify or verify the identity of individuals conducting in-person transactions for goods or services. This biometric is used to complement or replace authentication mechanisms such as presenting cards and photo identification, PINs, or signatures. Retail/ATM/point-of-sale applications can vary widely, depending on the level of supervision, the identifier used to claim an identity, and the existing authentication process. These applications can utilize either verification or identification, although the identification that takes place will normally be one-to-few: a search to identify an individual from a small subset of enrolled users. Retail/ATM/point-of-sale applications are similar to e-commerce/telephony, the other main customer-facing application, in that they are primarily voluntary: It is not in the best interests of institutions to compel customers to enroll in biometric systems. A primary difference between the two applications is that retail/ATM/point of sale is most often supervised during enrollment and verification.

Today's Retail/ATM/Point-of-Sale Applications

Biometrics are currently used in a small number of retail/ATM/point-of-sale applications, using a variety of technologies. While frequently modest in scope, these initial applications serve as effective proofs of concept, demonstrating that biometric usage in these environments is viable. Customers are familiar with the requirement for authentication at ATMs and points of sale, and merchants view biometrics as a way to reduce fraud and streamline the authentication process, while acting as a competitive differentiator.

ATMs and ATM-oriented multifunction kiosks are often viewed as being perfectly suited for biometrics, as a replacement for PINs and eventually for cards. The use of PINs at ATMs is perhaps the most familiar automated authentication process in existence. The weaknesses of this authentication method—that PINs are susceptible to compromise, easy to guess, and easy to forget—are evident to most everyone. Biometrics have been integrated into specially retrofitted ATMs, replacing the use of PINs, but in most cases still requiring a card or other identifier (see Figure 13.5).

In retail/POS applications, biometrics (specifically, finger-scan) are used in select locations to pay for transactions from prefunded accounts. These applications link a user's biometric data to user checking or credit card accounts, which are debited upon verification. A handful of grocery stores, fast-food restaurants, and cafeterias currently utilize this type of biometric solution in conjunction with standard payment systems. Major retailers have not yet adopted these systems, perhaps due to the perceived novelty of biometrics in this environment as well as the need to deploy biometric infrastructure at the point of transaction. These systems currently entail fairly minimal integration into existing customer-facing systems, instead acting as standalone systems with separate databases. Biometric transactions are logged separately and are often keyed into the existing register system.

Figure 13.5 A biometrically enabled ATM.

Future Retail/ATM/Point-of-Sale Trends

The retail/ATM/POS markets likely emerge as a result of merchants' and financial institutions' desire to reduce fraud and, to a lesser degree, to establish a competitive differentiator. Check and credit card fraud are major problems for retailers, who are generally held liable for "card not present" and chargeback fraud. The implementation of a biometric verification system, where a precondition of check usage is enrollment and linking with an active checking account, can significantly reduce fraud, primarily through deterrence. This type of offering can be positioned as a convenience to consumers, who would not be required to produce a driver's license or similar identification during the biometric payment process.

One of the key factors that may lead to increased biometric deployment in this environment is the ability to reach a large number of users through a modest number of devices. While the actual deployment of in-person transactional biometric devices may be complex, a single device is capable of processing hundreds of users per day. In addition, while more likely to happen outside of the United States, synergies between smart cards and biometrics may drive retail/ATM/POS growth. One of the challenges of breaking into the retail market is the need to implement and integrate a biometric infrastructure, including acquisition devices and back-end processing systems. Smart cards can store biometric data and verify that the card belongs to the cardholder, which is the primary problem faced in retail applications. In the United States, it is very unlikely that a comprehensive smart card infrastructure will be rolled out in the next few years; outside the United States, the development of a smart card/biometric infrastructure is a strong possibility.

A factor that could inhibit growth in retail/ATM/POS biometric usage is complexity of deployment. As opposed to remote transactions, where acquisition devices are either in-place telephones or those that will be increasingly in place over the next one to two years, in-person biometric transactions require that devices be deployed and integrated with current systems. Integrating biometrics into ATMs has traditionally been expensive, and the advent of lower-cost ATMs increases the relative costs of incorporating biometric functionality. At point of sale, integrating biometrics with existing technology may be highly complicated and require customization, especially given the broad range of terminals, computer systems, and register systems with which the biometric may need to interact.

In-person biometric usage also requires training of enrollers, daily operators, and users. While a motivated user group will be capable of using biometric systems fairly easily, unmotivated users—such as those who might be enrolled in a mandatory system—may struggle with system usage.

Fear of system errors may also inhibit the use of biometrics in this area. Biometric systems are subject to false matching and nonmatching; in a retail environment, the repercussions of false nonmatching can be severe, as valid customers have their identity questioned. Policies must be established that balance the need to detect and deter fraud attempts with the awareness of the problems engendered by false nonmatching.

Retail/ATM/POS: Biometric Solution Matrix

Although biometrics can be a strong solution for retail/ATM/POS usage, there are some limiting factors on this biometric application, as indicated in the solution matrix (see Figure 13.6).

Exclusivity (5/10). A number of methods are used to perform authentication in retail/ATM/POS environments. Biometrics would need to uproot what is a flawed but very pervasive and well understood system of PINs, photo identification, and signatures. Though biometrics may solve the authentication problem more effectively than other authentication methods, it is not the only option.

Effectiveness (7/10). Biometric solutions are reasonably capable of performing well in retail/ATM/POS environments, though there may be questions about the logistics of incorporating biometric authentication into the existing process flow of in-person transactional authentication. Some solutions would be highly effective, while others may be subject to high levels of false rejection. The effectiveness of biometrics in these environments can also be hindered by the relative infrequency of authentication—a user might interact with a point-of-sale biometric system only on rare occasions, meaning that the biometric data and the method of presentation may have changed over time. Since merchants cannot afford to inconvenience in-person customers, effective operation is essential.

Receptiveness (8/10). Although merchants will be more motivated to integrate biometrics into customer-facing applications because fraud reduction benefits them most, consumers have responded positively to the introduction of opt-in biometrics in these environments. The broad understanding of the need for authentication in this environment adds to the receptiveness expressed toward biometric solutions; the only factor that might impact receptiveness would be an increase in mandatory deployments.

Urgency (5/10). The problem of authentication in in-person transactions is urgent because reasonably effective authentication methods are in place, and policies and procedures have been developed over time through which potentially fraudulent transactions can be flagged. This is an impediment for biometrics to overcome in this area—while authentication is an omnipresent issue in retail/ATM/point-of-sale environments, it is not an urgent problem. At ATMs, for example, the maximum amount that can be stolen through use of an appropriated card is $300, which may not be enough to warrant introduction of a technology that could on occasion reject authorized users.

Scope (9/10). The potential scope of biometric authentication opportunities for in-person transactions is vast. Nearly everyone transacts on a regular basis in a retail/ATM/POS environment, meaning that biometrics could reach a very large number of customers. The scope represents an opportunity and a challenge. There are so many points of in-person transactional authentication that tremendous deployment effort would be necessary before reaching even a small percentage of locations. Furthermore, reaching a critical mass of enrollees would likely take several months, if not years.

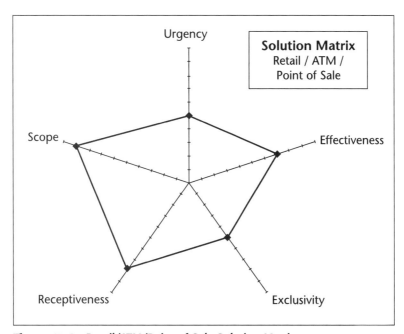

Figure 13.6 Retail/ATM/Point-of-Sale Solution Matrix.

Related Biometric Technologies and Vertical Markets

Facial-scan, finger-scan, and iris-scan have been deployed successfully at ATMs and expanded ATM-based kiosks in North America, Europe, and Asia. However, there is not yet a typical ATM transaction. Typical usage includes facial-scan verification at check-cashing kiosks, in which the system requires users to enter their social security number as a unique identifier. The facial-scan component is used to verify claimed identity as well as for 1:N identification against a derogatory database. Finger-scan has been used at credit union ATMs—customers are required to present a card or identifier to enable a 1:1 match with the finger-scan technology. Iris-scan, the technology most capable of performing 1:N identification, has been used in ATMs in identification mode—the user is identified without entering any data or presenting a token. Of the three technologies, facial-scan has been the most successful in ATM/kiosk deployments, with over one million users enrolled, but the biometric plays the smallest role. If the facial-scan component fails to match a user, which may be a fairly frequent process given the verification environments, users are routed to a live operator to provide nonbiometric verification.

Finger-scan is likely to be the primary technology used in retail and point-of-sale solutions, as it provides a solid balance between accuracy and ease of use. Facial-scan technology requires improvement in accuracy in order to be reliably used for more than derogatory database comparison. Iris-scan, while an accurate solution, may be a more expensive solution than can be accommodated in these environments.

Cost to Deploy Biometrics in Retail/ATM/POS

No firm benchmarks have been established for deploying biometrics in retail/ATM/POS environments. At the low end, finger-scan devices priced at approximately $500 are robust enough to withstand heavy usage in retail checkout counters. The most expensive iris-scan ATM systems in the late 1990s added tens of thousands of dollars to the cost of ATMs. For the most part, the costs of retail/ATM/POS deployment occur in integration into current systems and can vary widely according to the scale of the deployment and the complexity of existing systems. The least expensive solutions would be stand-alone biometric systems not integrated into existing authentication systems, which require that cashiers key biometric transactions into the existing register system. These systems might be leased on an ongoing basis, and costs to deployers would likely be a monthly flat fee plus a small percentage of biometric transaction value. Annual losses attributable to check fraud are estimated to be in the billions of dollars; any technology capable of reducing this amount, or establishing a framework within which legitimate customers can voluntarily transact, will be worthwhile to retailers. The implementation, integration, and training elements of in-person authentication systems are generally incorporated into the maintenance fee, in an effort to reduce the up-front expenditures.

One of the reasons that this sector is appealing to biometric companies is its transactional revenue model. Biometric hardware manufacturers and algorithm developers will not derive a great deal of revenues from these systems—a handful of devices would be capable of addressing the needs of a large store. Instead, the companies that enable biometrics as a transactional process and offer full biometric payment systems solutions will be the beneficiaries.

Issues Involved in Deployment

The following are the primary issues deployers must address when designing retail/ATM/POS systems.

Process Flow and Ergonomics

Users are accustomed to certain authentication processes in retail/ATM/POS systems. Deployers need to be sure that the new authentication method introduced has a straightforward user interface and process flow, and is easy for nonacclimated customers to use. While the biometric system should be positioned as a convenience tool that provides the additional benefit of security to deployers, if the system proves to be cumbersome to use during enrollment or verification it will be viewed as a hindrance. Because the accuracy of biometric systems is often based on users' ability to interact with the devices in a consistent fashion, ergonomic layout of acquisition devices is a similarly important consideration for deployers.

Integration into Existing Systems

Deployers must determine the appropriate level of integration of biometrics into existing systems. For ATMs, tight integration is required to allow successful verification to trigger cash dispersal. There are more decisions to be made in retail and point-of-sale environments. Systems can be implemented that tie directly into cash register systems, such that biometric authentication directly results in a transaction being executed. On the other hand, standalone systems, less expensive but with narrower functionality, can be implemented that require manual linking of biometric and nonbiometric systems. For retailers planning to implement customer-facing biometrics on a large scale, tight integration is more desirable because it enables more robust auditing and tracking of biometric transactions.

Motivation for Customers

The primary beneficiaries of biometric deployments in retail/ATM/POS environments are institutions, not customers. Customers may appreciate not having to present identification or remember PINs, but in most cases they will still need to use cards or other identifiers in order to transact biometrically. In some environments, customers may be required to enroll in a biometric system in order to gain special privileges such as check writing. Institutions must find a way to motivate customers to enroll in these systems in order to reach a critical mass of users. Institutions must also explain to users for whom biometrics are made a precondition of check writing (for example) why the system is being implemented and why it is not privacy invasive.

Technology Obsolescence

In large-scale retail/ATM/POS applications, deployers must ensure that the technology and devices implemented have a long life span. The time required to reach a critical mass of enrolled users could be substantial; since replacing biometric systems normally requires reenrolling users, systems must be capable of operating for an extended period of time from enrollment.

Conclusion

Biometrics are currently a very small niche solution in retail/ATM/POS applications. Whether the technology emerges from this niche status to become a common part of in-person authentication or remains an infrequently implemented solution depends on whether the cost and complexity of deployment are outweighed by increased security, convenience, and fraud deterrence.

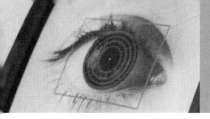

CHAPTER 14

Biometric Vertical Markets

A large percentage of biometric deployments occur within five key vertical markets. These market sectors may be characterized by an inherent need for user identification or authentication or by the presence of valuable or sensitive data. In these vertical markets, biometrics can serve to increase security or reduce fraud, and they may provide functionality not achievable through other authentication methods.

Deployers should be wary, however, of overemphasizing the vertical nature of biometric implementations. While financial services and healthcare, to cite two examples, are industries in which biometrics are being used successfully, there may not be anything distinctive about these deployments in these markets. For example, a hand-scan unit controlling employee access to a secure room performs identical functions whether it is deployed in a bank, a warehouse, or an airport. Similarly, the challenge of protecting a Windows network with a biometric peripheral does not change when moving from a financial sector to a corrections environment. Though a deployment might occur within a vertical, it does not always follow that the nature of the deployment was determined or informed by the vertical.

Five Primary Biometric Vertical Markets

While almost all biometric deployments can be categorized according to technology and application, as discussed in previous chapters, vertical analysis cannot account for all biometric sales. A fair percentage of biometric usage cannot be classified according to a particular market. Because of this, discussion of five key verticals is a more reasonable approach.

Law enforcement. The law enforcement market includes the use of biometrics to identify or verify the identity of individuals (1) apprehended or incarcerated because of criminal activity, (2) suspected of criminal activity, or (3) whose movement is restricted as a result of criminal activity. The biometric may be used to identify noncooperative or unknown subjects, to ensure that the correct inmates are released, or to verify that users under home arrest are in compliance.

Government sector. The government-sector market includes the use of biometrics to identify or verify the identity of individuals (1) interacting with a government entity or (2) acting in the capacity of a government employee or official. Military use of biometrics is best viewed as government-sector usage (law enforcement usage is a separate classification). This market includes usage such as individuals providing biometric data to an applicant processing system (forwarded to the FBI for background checks), a government employee accessing an internal network, and applicants for state welfare benefits interacting with a civil AFIS system.

Financial sector. The financial-sector market includes the use of biometrics to identify or verify the identity of individuals (1) interacting with a financial-sector entity or (2) acting in the capacity of a financial-sector employee. This market includes usage such as individuals accessing a bank account through a biometrically protected telephony system and a bank employee accessing biometrically protected internal networks.

Healthcare. The healthcare market includes the use of biometrics to identify or verify the identity of individuals (1) interacting with a healthcare entity or (2) acting in the capacity of a healthcare employee or professional. This market includes usage such as patients accessing medical information online through biometric authentication, or a hospital employee entering a room secured by a biometric.

Travel and immigration. The travel and immigration market includes the use of biometrics to identify or verify the identity of individuals (1) interacting, during the course of travel, with a travel or immigration entity or (2) acting in the capacity of a travel or immigration employee. This market includes usage such as an individual being authenticated by a ticket-issuing kiosk, an individual providing biometric authentication to a checkpoint, or an employee entering a secured area of an airport.

Law Enforcement

The law enforcement market is characterized by the widespread use of AFIS technology, implemented at the state and federal levels across the United States and, increasingly, around the world. The AFIS market is mature by biometric standards, having already reached a substantial percentage of its potential deployment areas. Developments in the AFIS market include new portable devices tied to central databases for field suspect identification. These devices will find adoption in jurisdictions ready to complement first-generation AFIS technology.

Aside from AFIS, the law enforcement market is characterized by the increased use of automated mug shot processing applications. Facial-scan technology can leverage existing acquisition technology and databases to locate possible duplicate identities for arrested subjects. The law enforcement market may also be significantly changed by the possible emergence of DNA as an automated biometric. DNA matching, while not yet automated to the point where it can be considered a true biometric technology, may eventually be used in a range of law enforcement applications.

Technologies Used in Law Enforcement

AFIS technology is clearly the central technology in the law enforcement space, due to the prevalence of fingerprints stored at the state and federal levels. Facial-scan, by virtue of its ability to search mug shots and provide surveillance functionality, is the other technology essential to this market's future. It is possible that voice-scan and iris-scan will also gain some percentage of the overall law enforcement sector, but these technologies are highly application-specific and should only find niche usage.

Typical Law Enforcement Deployments

Within the law enforcement market, biometrics are deployed in AFIS, surveillance, mug shot processing, corrections, and probation/home incarceration applications.

AFIS

AFIS deployments are ubiquitous in local, state, and federal law enforcement applications—common enough that a discussion or analysis of individual deployments is beyond the scope of this report. Local police stations are major purchasers and deployers of AFIS devices used to capture and transmit fingerprints from arrested suspects. The multimillion-dollar AFIS matching systems, which process fingerprint data from suspects and crime scenes, are implemented at the state, regional, and national levels. The integrated FBI AFIS database, known as IAFIS, currently has 40 million 10-print records in its databases and conducts thousands of searches daily for matches on criminal suspects. A major shared AFIS network, known as Western Identification Network (WIN), also extends across the western half of the United States.

Surveillance

Though not broadly used today, surveillance is projected to become a major growth area for law enforcement applications. The first public deployment of surveillance technology in Newham, England, uses 144 closed-circuit television cameras to record all activity on several streets in what had been an unsafe neighborhood. The facial images acquired through the cameras are compared to a hotlist of known offenders. Police in Tampa Bay, Florida, use 36 cameras to acquire facial images from passersby in an entertainment district, comparing them to a hotlist and discarding the nonmatches. Any faces that come back as hits—indicating that there may be a match—are investigated manually. Similarly, the Royal Canadian Mounted Police use a system in the Pearson airport, in Ontario, to compare the surreptitiously videotaped faces of suspicious-looking individuals with a police database. Law enforcement use of surveillance technology is likely to grow rapidly despite recent controversy in the United States.

The potential for law enforcement-related surveillance applications is substantial, and it is highly likely that biometrics will be sold and licensed in the United States and abroad for incorporation into existing video capture systems. Deployers should be aware of some of the limitations of facial-scan technology in this area. The nature of surveillance applications is such that only a very small percentage of individuals are likely to be correctly identified. Variables such as lighting, camera angle, image quality, distance of the subject from the camera, and the size of the watch list database, to name a handful, make surveillance an extremely difficult application. A large number of users may be incorrectly flagged in an effort to find wanted suspects, while the suspects themselves may well go unidentified.

Mug Shots

Facial-scan is also used to compare mug shots against databases of known or wanted offenders, leveraging nonproprietary equipment for image capture. In August 2000, mug shot galleries in police departments in Kent County, Michigan, were augmented with facial-scan software in order to aid in future capture and identification of suspects. A number of municipal police departments in California have deployed similar systems. Mug shot searching is a near-ideal use of facial-scan: It does not involve the introduction of new acquisition devices, it automates what is currently a manual process, it occurs within a controlled acquisition environment, it performs rapid searches, and it is not relied upon to provide 100 percent accuracy.

Corrections

Finger-scan and iris-scan technologies are used in corrections applications to process incoming and outgoing inmates. While finger-scan is widely deployed in these environments, iris-scan deployments are less prominent. Prisons in Pennsylvania, New York, and Florida use iris-scan technology to verify the identity of convicts before release. This system is designed to avoid the possibility of accidentally releasing or granting privileges to the wrong prisoner as the result of fraudulent identity claims. Other implementations in correctional facilities enroll visitors to ensure that people leaving the facilities are visitors, not inmates. Corrections environments are challenging due to the potential for encountering a highly uncooperative user population.

Probation and Home Arrest

In addition to arrest and incarceration, biometrics are used for postrelease programs to ensure compliance with probation, parole, and home detention terms. While outside of facilities, people can be difficult to track, and such tracking can require extensive manpower and administrative resources. Typical methods of tracking, such as scheduled visits by officers or telephone calls, can result in a long delay between the time that a tracked individual disappears and the point at which the authorities become aware that he or she is missing. Voice-scan is an attractive solution to this problem: In order to verify that an individual is at home while under house arrest, an automatic system telephones the person at regular intervals. Using a challenge-and-response recorded dialogue, the system automatically verifies that the individual is at home through voice-scan. If the voice answering the phone does not match the voice template, or if no one answers the phone, the system immediately alerts officers so that appropriate action may be taken. Hand-scan technology has also been used in kiosk-based systems to verify the identity of probationary offenders, reducing the burden on probation officers.

Conclusion

Even if the commercial and private-sector biometric industries should somehow fail to emerge in the coming months and years, biometrics will continue to grow as a law enforcement solution. Biometrics are a fundamental part of law enforcement operations, and there is general acceptance of the role of biometrics in this area. Continued growth in this area does assume that the use of biometrics as a law enforcement tool will be circumscribed and that the technology will not be used for purposes for which the systems were not originally intended.

Government Sector

Interaction with government bodies is a constant part of most everyone's lives, whether in the capacity of benefits recipient, voter, taxpayer, employee, member of the armed forces, or business operator. In addition, individuals interact with governments both privately and in the course of employment, and may require different levels of authentication depending on the nature of a specific interaction. The very broad range of interaction between individuals and governments, as well as the need for strong authentication in many of these interactions, is the primary driver of the use of biometrics in the government sector.

However, the extremely broad potential for government-sector biometric usage does face challenges. It is possible that in the United States, Canada, and Europe—areas characterized by a heightened awareness of privacy concerns as well as a mistrust of government initiatives—biometrics will come to be viewed as an invasive technology when implemented by governments. The Big Brother fear is most likely to be invoked when governments are the deployers of biometrics—private-sector deployments are less likely to be mandatory, and law enforcement usage is taken as a necessity. Developing countries, as well as most developed Asian states, are less likely to react in this way to biometric usage, as their conceptions of privacy and acceptable levels of government interaction differ from Western conceptions.

Technologies Used in the Government Sector

A variety of technologies will be used in government-sector biometric applications. AFIS technology is the primary technology for locating duplicate enrollees in benefits systems; finger-scan is the primary technology for desktop and card-based ID solutions; facial-scan will likely find numerous implementations in image-based ID systems; and other technologies will find niche applications within which they can operate. In many cases, technologies are specifically developed for their efficacy in government-sector implementations.

Typical Government-Sector Deployments

There are several categories of government-sector biometric usage, including national IDs, voter IDs, driver's licenses, benefits distribution, employee authentication, and military usage.

National IDs

National citizen identification programs are the most ambitious biometric programs, often incorporating both 1:N identification and 1:1 verification in conjunction with large-scale card issuance programs. National and local ID cards requiring biometric sample capture prior to issuance of the card are used for multiple purposes: collection of benefits, voting privileges, drivers licenses, and so on. These technologies are based primarily on finger-scan and AFIS technology, though it is possible that facial-scan and iris-scan technology could be used in the future.

Many of these implementations are in early stages, as the issuance of new identification cards is usually a multiyear process and the logistics of enrollment, identity verification, card manufacturing, and duplicate resolution can be overwhelming. Many of the countries integrating biometrics into the ID issuance process are in the developing world; similar programs proposed in Europe and the United States have met with various objections. Note also that large-scale projects in developing countries are often delayed, revoked, contested, or altered in scope, and can change overnight.

Nigeria. The government of Nigeria entered a $30 million agreement to supply a national ID card to its citizens in late 1998. Card issuance will follow a 1:N search for duplicate identities and will be used for voter registration, among other functions.

Argentina. Argentina embarked on a national ID project in late 1998. This is expected to be one of the largest AFIS deployments worldwide, with the capacity to enroll 50 million citizens. The deployment involves conversion of existing records to searchable electronic format as well as the issuance of new identification papers.

Yemen. After a successful pilot in 1998, Yemen intended to deploy a national identity card in 2002. The card uses finger-scan biometrics for identification, and the program will ultimately scale to 15 million citizens.

China. Though in the very early stages, the Chinese government is piloting a smart card-based national ID system. The cards will be used for various government functions and will contain finger-scan templates. Since the biometric data is to be stored only on the cards and is not collected through high-resolution live-scan devices, the system will not provide 1:N functionality. Over 20,000 citizens are already participating in the program, which could eventually scale to over 800 million citizens.

There are countless issues involved in biometric scale national ID programs: the limits of large-scale biometric identification, the logistics of enrollment, initial identity establishment, maximum permissible response time, the link to card issuance, the privacy implications, fallback procedures, costs, likelihood of detection and deterrence, technology obsolescence, and reenrollment, to cite only the obvious. Determining whether and how to go about deploying biometrics in conjunction with a national ID program can be a substantial project in itself before the first user is enrolled.

Voter ID and Elections

Several countries have instituted biometric schemes for national or local elections, often in conjunction with card issuance. The rationale for biometrics in a voter registration program is almost always 1:N searches to locate multiple enrollees, although the technology can be used for 1:1 verification. The biometric disciplines utilized in this environment are facial-scan, finger-scan, and AFIS.

Mexico. The Mexican government has licensed facial-scan technology in order to find duplicate enrollments in its registered voter database. After its one-time usage in the July 2000 Mexican national elections, a permanent license for the technology was acquired. The purpose of the system is to deter and prevent citizens from voting more than once, using assumed names.

Peru. In 1997, Peru implemented a combined voter identification and national identity card project, with a database of finger-scan images captured at card issuance. Approximately 12 million citizens will be enrolled in the final iteration of the project.

Brazil, Dominican Republic, Costa Rica, Panama, and Italy. A handful of countries use biometrics for voter verification at polling stations. Brazilian elections utilize biometrics to facilitate electronic voting, a project begun in 1996. Finger-scan terminals were added to the voting kiosks to provide more robust personal authentication for voters. The system allows over 35 million citizens to vote directly on computers and is likely the single largest electronic voting system worldwide. Similar systems using finger-scan peripherals exist in the Dominican Republic, Costa Rica, Panama, and Italy.

Uganda. In early 2001, the Ugandan government implemented a facial-scan system for national elections in order to eliminate duplicate voting among its 10 million citizens. This deployment met with some objections—it was suggested that the technology was implemented as a means of intimidating opposition voters.

The use of biometrics for voter ID and elections can both detect and deter fraudulent activity, and can be especially valuable in countries that lack robust means of identification.

Driver's Licenses

Both facial-scan and finger-scan biometric solutions have been successfully deployed in driver's license programs in several U.S. states and in several other countries. Driver's license implementations are generally always 1:N, although jurisdictions where the driver's license is used as an identification card may also incorporate 1:1 functionality. Efforts are under way to standardize and share finger-scan data between U.S. states, as individuals barred from driver's license issuance in one state can, in the absence of biometric identification, simply acquire a license in a neighboring state.

United States. The Illinois driver's license system incorporates facial-scan technology for its 25 million residents, processing existing and newly collected images to search for likely duplicates. West Virginia's driver's license system also incorporates facial-scan technology, used for duplicate location as well as for the issuance of ID cards to minors in order to establish a database to help identify and locate missing children. Georgia's driver's license program collects finger-scan data for its 4 million residents, although the data is not yet used in any 1:N identification. Colorado and Texas also utilize digital fingerprint imaging technology.

Gujarat and New Delhi, India. Both vehicle registration and driver's license cards in New Delhi and Gujarat, India, incorporate biometric solutions. The project in Gujarat, using smart cards and containing a finger-scan template, is intended to scale to 10 million residents.

The use of biometrics to detect and deter duplicate identities in driver's license applications is especially appealing when issuing special-privilege licenses, such as those for transporting hazardous materials. In countries such as the United States, where licenses are controlled at the state as opposed to the federal level, there is also a need to share data between states to ensure that individuals cannot circumvent the system by simply applying in an adjoining state. This standardization is a major challenge for jurisdictions, as vendors' biometric systems are proprietary and the data by and large cannot be interchanged.

Benefits Distribution

Identification systems used for disbursement of government benefits require advanced security to protect against fraud and abuse. Government benefits programs provide attractive targets for fraudulent activity; abuses can result in increased spending on government programs and can siphon away funds necessary for legitimate citizen benefits. Along with the ability to alleviate large amounts of fraud and abuse, biometric solutions can provide quick and convenient access to benefits, ensuring that legitimate recipients are able to receive the services they need in a timely manner.

Spain. Spain's Tarjeta de Afiliación a la Seguridad Social (TASS) program is an initiative established in 1994 to provide smart cards for the recipients of healthcare, social security, and unemployment benefits. The cards store a finger-scan biometric template locally for user verification—no central database is maintained, and the system does not provide 1:N functionality. The purpose of the card and the biometric system is to enhance citizen access to services while protecting the security of their personal information and protecting the cards from misuse.

Philippines. The Republic of the Philippines engaged in a six-year program beginning in 1998 to distribute national social security cards. The program is expected to reach a total of 35 million recipients. The cards contain a finger-scan biometric template compliant with AFIS technology. The card is designed to prevent fraud and to facilitate secure distribution of benefits to legitimate cardholders, and offers both 1:N and 1:1 transactional functionality. The card combines one- and two-dimensional barcodes as well as a magnetic stripe and other security features.

South Africa. In order to eliminate corruption in South Africa's pension distribution system, the South African government awarded a contract for the issuance of a finger-scan-enabled smart card for pension recipients. Prior to the implementation of this card, benefits distributors would simply steal, at point of issuance, cash disbursements intended for the recipient. Currently deployed to three million South Africans, the card has significantly improved the lives of recipients, many of whom are illiterate and without other means of identification.

United States. The U.S. welfare system, a common target for fraud and abuse, has deployed biometric solutions to eliminate double-dippers in Arizona, Texas, California, New York, Connecticut, and Massachusetts. Though cost avoidance attributable to fraud detection and deterrence is always difficult to pinpoint—external factors such as economic trends and other welfare reform initiatives impact fraud detection and deterrence—most states have reported positive results from their welfare avoidance programs.

In the United States, portrayals of biometrics deployed in benefits distribution programs often emphasize what is seen as a punitive element: Benefits recipients, the poorest classes, are subject to the same type of treatment as criminals. However, it is worth noting that in developing countries, where individual identity is more difficult to establish than in the United States for various cultural reasons, biometrics are often viewed as an empowering technology, ensuring that only valid recipients can receive benefits. Without biometrics, distribution can be highly ineffective and fraud ridden.

Employee Authentication

Although employee-facing biometric applications are not as high profile or controversial as citizen-facing applications, government use of biometrics for PC, network, and data access is significant. There are hundreds of deployments in this arena, along the lines of the following examples.

International legislatures. In 1999, the 520-seat Mexican congress installed a finger-scan system for authorizing congressional votes. The 290-seat Venezuelan congress installed a similar finger-scan system for authorizing votes in 1997. The 568-seat Turkish parliament also uses finger-scan to authorize legislative votes.

Oceanside, California. A finger-scan solution was employed to secure a government network in Oceanside City, California, in 1999. After a successful pilot in which the devices were tested on 50 seats out of the 1,700-seat network, the city outfitted 1,500 of the network workstations with finger-scan peripherals and software.

Glendale, California. The city of Glendale, California, is implementing finger-scan peripherals and accompanying software to all 2,100 city employees, with initial rollout to the Glendale Police Department. The benefits are increased network security, as well an estimated savings of $50,000 per year on administrative costs due to the increased efficiency and convenience of network access.

U.S. GSA Common Access Card. The General Services Administration Common Access Card is a multiyear, $1.5 billion program to replace U.S. government employee and military identification cards with smart cards. The cards will be designed for use in physical access, logical access, and identification, and are meant to be interoperable between departments. Biometrics are not a mandatory feature, but many agencies will opt for biometric storage on the card-based chip for future usage in network and building security.

Military Programs

For several years, the U.S. military has been conducting tests on the applicability of biometric solutions for various functions. A notable example was in Fort Sill, Oklahoma, in which the U.S. military and Treasury piloted a salary distribution system using smart cards, kiosks, and finger-scan technology. The cards were multifunctional and enabled purchases in local retail establishments. The pilot was conducted from 1998 to 1999, with 18,500 recruits participating. A total of $4 million in salary was disbursed through the smart cards.

In more traditional military applications, governments are investigating the use of biometrics for personnel identification in hostile environments. Uses may include authenticating individuals allowed to operate weaponry or machinery and identifying wounded personnel. Because the operational environment for military usage is much more challenging than in normal biometric implementations, substantial R&D will be necessary before today's technologies are suitable for use in military environments.

In the United States, a centralized Biometrics Management Office has been established to "lead, consolidate and coordinate all biometric information assurance programs of the DOD in support of Network Centric Warfare."

Conclusion

In the wake of September 11th, it is even more likely that government-sector biometric deployments will be a major growth area for biometrics. Preliminary discussions about a national ID — whether legally mandated or emerging from a standardized driver's license — have also arisen, most of which include the use of biometrics for verification or identification.

Financial Sector

Banking and financial services represent enormous growth areas for biometric technology, with many deployments currently functioning and pilot projects announced frequently. The varied authentication requirements of financial institutions create opportunities for diverse implementations for both customer-facing and employee-facing authentication. Biometrics will not only secure and improve existing services but may enable new, high-value remote services that would not otherwise be possible.

The finalization of standards in biometric data formats and encryption, including BioAPI, BAPI, CBEFF, and X9.84, will help drive acceptance of biometric technologies in an industry reluctant to deploy unfamiliar technology. The continued emergence of alternative payment platforms for Internet commerce, in which a degree of anonymity can be maintained, will also leverage biometric technologies. A major variable is whether large transactional authentication companies—credit card providers, as well as companies that provide the infrastructures for check and credit clearance—will incorporate biometric authentication within their infrastructures. It is highly likely that these major players will implement some type of customer-facing biometric solution.

In addition, efforts are under way to investigate the use of biometrics in developing economies where authentication methods are much less advanced than in Westernized countries. Though years may be required to develop such offerings, this represents a giant untapped market for biometric revenues in financial services.

Legislative restrictions on the use of common identifiers, such as those found in the Gramm-Leach-Bliley Act, may also prompt the use of biometrics as semi-unique identifiers that cannot be traced across disparate databases. As opposed to social security numbers, which are static and can function as common identifiers, biometric data changes from verification to verification, making it less likely to be used to link personal information.

Growth of biometrics in the financial sectors has been limited by the conservatism of many financial institutions. In particular, the risks involved in customer-facing applications, such as fraudulent account access and denial of account access to legitimate customers, have compelled financial institutions to retain their current authentication processes. These current processes have the benefit of being known quantities to financial institutions. In addition, years of mergers between major financial service companies have resulted in companies with dozens of legacy systems to manage. Integrating biometric functionality into these back-end systems may be a major challenge, as biometric technology can be comfortably integrated only within a handful of platforms.

Technologies Used in the Financial Sector

As of early 2001, finger-scan, facial-scan, iris-scan, retina-scan, signature-scan, voice-scan, and hand-scan have all been deployed successfully by financial institutions worldwide for transaction authentication, access to computers and networks, and physical access. The varied nature of financial-sector deployments means that no single technology will dominate this segment, although the need for accuracy in most financial-sector deployments suggests a strong role for finger-scan and iris-scan. On the other hand, facial-scan has processed more customers than any other technology in this sector, and voice-scan will certainly be deployed in a range of telephony applications.

Typical Financial-Sector Deployments

Financial-sector deployments incorporate both customers and employees, and range from internal deployment to remote and kiosk-based authentication.

Account Access

The additional security offered by biometric systems, along with the ability to keep definitive and auditable records of account access by employees and customers, has resulted in deployments of various technologies for transactions within banks. Westernbank in Puerto Rico is notable for its multiple biometric deployments. Using finger-scan, customers can access accounts and employees can log into their workstations in all 37 Westernbank branches. Bank customers also have the option of using a signature-scan system for verification. Since the Westernbank systems are intended primarily for customer convenience, biometrics use is available as an option to all customers who wish to use the system. Westernbank representatives stated in mid-2000 that of approximately 300,000 bank members, 10 to 15 percent opt to use the biometric systems and the rest prefer to use traditional technologies. Westernbank estimates the total cost of the deployment to be approximately $3 million. In mid-2000, Charles Schwab & Co. began to allow customers elective use of a signature-scan system for new account applications. Primarily for customer convenience, the system is designed to maximize the efficiency of transactions.

ATMs

For certain types of transactions, biometrics can be an effective ATM solution, but the long-term viability of biometrics in this environment is uncertain. Iris-scan was one of the first biometric technologies to have been piloted in an ATM environment. In 1997, Nationwide Building Society, a savings and loan institution in Swindon, England, successfully piloted a cardless iris-scan ATM. The pilot won high approval ratings—94 percent of customers preferred iris-scan to PINs. Similar customer-facing pilots took place at Bank United in Texas and Takefuji Bank in Japan. In Frankfurt, Germany, Dresdner Bank has been conducting an iris-scan ATM pilot since late 1999. Thus far, only bank employees have been given the opportunity to use the system, with about 1,000 enrolled as of late 2000.

Finger-scan technology has also been deployed in ATMs with some success. Purdue Credit Union in Indiana installed a series of kiosks on Purdue University campuses that allowed members to conduct banking transactions using finger-scan authentication. The project began in May 1997 and was fully functional by June 1999. Members of the Houston Municipal Credit Union also have the option of using finger-scan at kiosks for transactional authentication.

Since transactions executed at ATMs are generally not large, the benefits of biometric authentication—and the risks of retaining the PIN/card paradigm—are not overwhelming. User convenience may be enhanced through biometrics, but this is tempered by the risk that users will not perceive the technology positively. The lack of a perceived need for heightened security, coupled by the risk that user convenience may not outweigh user prejudices against biometrics, may limit incentives to deploy biometrics in ATMs.

Expanded-Service Kiosks

A more receptive market for biometrics may be special-purpose kiosks, using biometric verification to allow a greater variety of financial transactions than are currently available through standard ATMs. InnoVentry deployed over 1,000 Rapid Pay Machine (RPM) kiosks across the United States, mostly in convenience and grocery stores, enabled by facial-scan technology. The kiosks use facial-scan for customer verification and to search against derogatory databases. After enrollment in the system, involving submission of credit-related data, the RPM cashes payroll and government checks. The system is expected to incorporate new functionality, including bill payment, wire transfers, and other banking services. Initially targeted at self-banking customers without bank accounts, the kiosks are expanding to provide services to people with bank accounts. Over 1 million users have enrolled in the InnoVentry system, making it one of the largest private-sector deployments of biometrics in the world.

In Mexico, Groupo Financiero Banorte, a leading financial institution, has deployed a system of kiosks utilizing smart cards and biometrics for workers who do not hold bank accounts. Using finger-scan technology, the smart card system obviates the need for paychecks and instead dispenses cash to employees. The smart card also functions as a debit card in local retail establishments. Some 4,000 workers participated in the pilot project, and the full deployment is intended to serve over 650,000 Mexican workers.

Online Banking

Internet-based account access is expected to grow into a consumer biometric application, but to this point it has been tested largely in pilot programs involving bank personnel. Because there is no biometric infrastructure to leverage, any implementation of biometrics in customer-facing Internet authentication requires that the institution supply readers to customers. In addition, the current PIN/password paradigm involved in online account access operates well enough that institutions have not rushed to complement or replace it. In November 2000, four Utah credit unions began a pilot using peripheral finger-scan technology. Employees use the device for authentication when conducting online transactions from home; pending results of the pilot, the credit unions plan to allow credit union members to start using the new system for private online transactions.

Telephone Transactions

Biometric voice-scan is used to enable secure telephone-based financial transactions. Voice-scan's unique advantage is the ability to leverage land and mobile telephones. The potential convenience of voice-scan telephone transactions is enormous, as bank customers can conceivably execute financial transactions from almost anywhere. Challenges for voice-scan deployments include negative consumer perception of voice-scan technology, as many people do not trust the ability of a voice-scan system to perform reliably.

PC/Network Access

Logical access is an obvious area of growth for biometric technology in many different industries. Largely using finger-scan peripheral devices, several financial institutions have engaged in pilots and deployments of biometrics for workstation and network access. These applications generally aim to increase network security and employee productivity. California Commerce Bank deployed a finger-scan system to over 200 employees in January 2001. The employees use peripherals to access customer accounts, applications, and network resources. The deployment was a response to recent legislation, the Financial Services Modernization Act of 1999, which mandated higher security standards to protect transaction integrity and customer privacy.

Physical Access

Access control is critical to financial institutions with on-site vaults, safe deposit boxes, and other areas in need of security. A major deployment is under way in Zion's Supermarket Banks in Utah, in which access to 475 safe deposit boxes is controlled with an integrated finger-scan device. The core benefit of the system is customer convenience as opposed to security, since the system allows customers to access their safe deposit boxes without needing to wait for the assistance of a staff member. A similar application of biometric technology exists in Germany, where nine different banks control access to safe deposit boxes using facial-scan technology. In April 2000, E*Trade, an online investment service, implemented a system to control access to secure areas using combination proximity card readers and standalone finger-scan devices.

Conclusion

Because of the widely acknowledged need for authentication in a range of employee- and customer-facing environments, the financial sector is seen as a major growth area for different types of biometric solutions. The emergence of biometric standards in the financial sector should help minimize the risks associated with biometric deployments in these sensitive applications.

Healthcare

Legislative requirements, a heightened awareness of patient privacy, and the increased availability of personal medical information online are expected to drive biometric revenues in this vertical. Areas of increased biometric revenue generation in healthcare will include protection of patient data on internal networks, customer-facing applications, and development of biometric software specifically for the healthcare market.

Healthcare has unique challenges and requirements that can be met with strong verification technologies. Recent legislation creates a demand for strong verification in new areas as part of an overall effort to improve the quality and efficiency of the healthcare industry. Biometric technology, being the technology with the strongest ability to reliably identify and verify individuals, is likely to be the most appropriate technology for fulfilling many unique requirements of the healthcare industry.

The Health Insurance Portability and Accountability Act (HIPAA), in particular, will drive biometric revenues in the healthcare vertical, although not overnight. HIPAA calls for standardization of medical data, unique identification for healthcare providers and recipients, and security standards for protecting health information. The third of these elements—security—will have the most direct impact on the biometric market. Biometrics are specifically cited in HIPAA's "Technical Security Services" as one of several means of *entity authentication* (along with password, PIN, telephone callback, token) from which healthcare providers must choose to protect healthcare data. Though not mandatory, biometrics are a means for healthcare providers to come into compliance with HIPAA regulations. HIPAA's impact on the biometric industry will be gradual, as healthcare organizations eventually become comfortable with the idea of using biometrics to protect electronic data.

Although HIPAA should be a driver for biometric revenues in healthcare, the entire information security industry is focused on delivering various types of HIPAA solutions. This means that biometric technologies will be competing with, or playing a secondary role to, technologies such as SSO and tokens. Proprietary medical applications may also pose a challenge to some biometric technologies. As in financial services, the healthcare providers may utilize legacy technologies not amenable to the incorporation of biometric technology.

Technologies Used in Healthcare

Finger-scan has traditionally been the most frequently deployed biometric technology, though there have been only a handful of publicly announced deployments in this sector. Because most biometric implementations in healthcare will involve access to data on networks and PCs, finger-scan will likely be the technology most commonly implemented. Finger-scan is well suited to desktop environments, providing a strong balance between accuracy and convenience. In addition, as remote access to personal data becomes common, finger-scan technology embedded in keyboards and peripherals will be used to secure online access. Iris-scan technology, assuming it is able to penetrate some portion of the desktop market, may generate substantial revenue in this sector as well.

Biometrics will also be used in kiosk-based healthcare implementations. Technologies suited for kiosk-based authentication, which may entail access to information or to disbursement of prescriptions, include finger-scan, iris-scan, and hand-scan. For basic access control applications, hand-scan will likely continue to garner some share of the healthcare market. However, it is possible that iris-scan technology, because it provides hands-free access and higher security, may be used for access control implementations as well. These implementations could include authentication of personnel entering controlled rooms or storage areas.

Typical Healthcare Deployments

Healthcare deployments are largely employee-facing, although it is anticipated that customers—in the form of patients and other healthcare recipients—will use biometrics to access personal data and to verify identity in conjunction with treatment. Many of the larger employee-facing network security deployments of biometrics have occurred in the healthcare sector.

PC/Network Access

A number of early-phase projects are currently in place that secure employee access to hospital networks. Most implementations of biometrics for network security will involve locking workstations, providing strong audit trails of user access to information, and integration into custom applications. Over time, biometrics will secure hospital VPNs, increasing the amount of data accessible to medical professionals from remote locations.

Since 1998, the New York State Office of Mental Health has purchased and deployed over 6,000 finger-scan units to enhance network security. The systems, used by healthcare staff, are driven in part by HIPAA legislation that demands increased measures to protect the confidentiality of sensitive patient information. St. Vincent Hospitals and Health Care Centers embarked on a 5,000-seat network security deployment in March 2001, using biometrics to authenticate single sign-on users.

Access to Personal Information

Over the next several years, the largest segment of biometric usage in healthcare environments may be enhancing individuals' access to their personal information. It seems likely that consumers will demand access to their information—one of HIPAA's principles—while demanding that such information be kept secure. Biometrics are well suited to serve this market segment. Medical information may be stored on smart cards or secured networks; in either case, the biometric is a gateway to personal information.

The most visible deployment of biometrics to enable access to personal information is actually one of the longest-running implementations of biometric technology: Spain's Tarjeta de Afiliación a la Seguridad Social (TASS) project, previously described under *Government Sector*. Using hundreds of kiosks across Spain, TASS allows enrolled individuals to access their personal medical information stored on a biometrically secured smart card. Treatments are approved and applications for medical benefits are processed through the network of kiosks. Though TASS is as much a kiosk and smart card project as it is a biometric project, biometrics are a necessary security piece without which the project could not have existed.

Patient Identification

In emergency medical situations, the need for conclusive authentication is paramount. There have been surprisingly few efforts to deploy biometrics as a solution to this problem. For individuals without identification and unable to communicate, biometric identification provides unique capabilities. Challenges involved in devising this type of biometric identification system include enrolling a sufficient number of users to achieve critical mass and the availability of biometric devices in emergency situations and locations.

AFIS technology has been implemented in two hospitals in Indiana and Kentucky for the purpose of patient identification. In response to a situation in which victims of a shooting could not be identified, the opt-in systems are designed to identify users from databases of approximately 10,000 to 15,000 individuals.

Conclusion

While biometrics are already being seen as a viable health care solution, they will compete with other authentication and security technologies cited under HIPAA legislation. In some cases, biometrics may be implemented in conjunction with other authentication methods.

Travel and Immigration

The range of potential applications for biometrics in travel and immigration environments contributes to its rapid growth rate. Prominent deployments of biometrics in travel and immigration include airport employee access control, passenger control in domestic and international air travel, and identity verification at international borders. Security is a major concern in travel and immigration environments: There is reduced expectation of privacy and anonymity and a generally acknowledged benefit in strong authentication. At the same time, the movement of passengers cannot be unduly impeded. Biometrics are becoming more frequently deployed as a solution to this two-sided problem, providing high levels of security to airport and customs officials while offering convenience for passengers.

In addition, there are a number of potential applications of biometrics in travel and immigration environments unrelated to the core business of moving individuals. It is reasonable to expect, for example, that biometrics will be deployed as a PC/network access solution for employees.

One of the challenges facing biometrics in travel and immigration applications is that biometrics are not idealized for strong performance when used infrequently. Desktop biometric applications, in which an employee might verify dozens of times a day, reinforce proper usage through repetition and habituation to the verification process. In certain travel environments, especially those involving passengers or customers, the relative infrequency of device usage may reduce system accuracy. If biometric solutions reject authorized travelers with any degree of frequency, they will meet with substantial objections.

Technologies Used in Travel and Immigration

Hand-scan, finger-scan, and iris-scan are the technologies most likely to be deployed in travel and immigration environments through 2005. Currently, hand-scan is the technology most frequently deployed, used for employee access control and passenger customs processing. Iris-scan technology is being piloted in employee and passenger-facing implementations; it provides a higher degree of accuracy while allowing hands-free operation. Because iris-scan projects are still in the early stages of deployment, their performance over time and acceptability to the general populace remain to be seen. Finger-scan technology has found limited deployment in this subsegment, although it is a strong contender to find increased deployments in travel and immigration between now and 2005. In particular, the use of finger-scan in large-scale passport programs is expected to increase substantially, as governments opt for 1:N functionality in addition to 1:1 functionality.

AFIS acquisition and matching technology is also used in travel and immigration to perform employee background checks. Applicants are fingerprinted during the hiring process, and their data is submitted to the FBI for processing. The fingerprints are matched against a felony database in order to ensure that the applicant's identity claim is valid. Live-scan optical fingerprint systems, as opposed to ink cards, are increasingly used in this type of applicant processing. AFIS growth in this segment is expected to remain steady.

Typical Travel and Immigration Deployments

Travel and immigration deployments can incorporate citizens, employees, or customers (in the form of travelers), with leading applications including air travel, border crossings, employee access, and passports. Many substantial employee-facing deployments of biometrics have occurred in the travel and immigration sector.

Air Travel

Biometric systems are used by frequent travelers to circumvent customs and immigration lines. Users undergo a thorough screening process for program entry and on subsequent crossings use biometric authentication without interacting with immigration personnel.

Travelers entering the U.S. can enroll in INSPASS, circumventing waiting lines in the airport. The system, based on hand-scan technology, has been in place since 1993. In 1998, Ben-Gurion Airport in Tel Aviv, Israel, installed a similar system to allow Israeli citizens to circumvent lines when traveling internationally. The implementation was recently expanded due to increased passenger demand, processing 50,000 passengers per month, making it one of the most successful biometric projects in this sector.

An ambitious implementation involving passenger processing began in July 2001 at London's Heathrow Airport. This implementation allows selected frequent travelers on British Airways and Virgin Airways to clear immigrations through iris-scan verification. This is the first use of iris-scan technology in a passenger-facing implementation. A similar project was announced in Canada, under Canadian Airports Council Expedited Passenger Processing System Project. Frequent passengers at eight major Canadian airports may elect to use travel cards storing a biometric template. Cards will allow the passengers to verify their identity at unattended kiosks to simplify border crossings.

Border Crossing

The use of biometrics to authenticate and identify individuals at national or state borders is expected to increase, as a number of high-profile projects have been launched over the past several years. The government of Israel, seeking to alleviate long waiting lines and crowds at border crossings between the Gaza Strip and other areas of Israel, instituted the Basel project in 1998. This implementation, when finalized, will utilize biometrics and proximity cards to facilitate passage through the border checkpoint. The implementation will use hand-scan and facial-scan; the combination of technologies is designed to decrease processing time and reduce false nonmatches. Expected to process tens of thousands of Palestinian laborers per day, this deployment is particularly challenging due to a combination of high security requirements, a high volume of subjects, and ingress and egress issues.

The U.S. government, in conjunction with the governments of Canada and Mexico, has participated in several pilot programs testing the feasibility of biometric deployments across American land borders. The southern border is particularly problematic for border control because of illegal immigrants and unauthorized foreign workers entering from Mexico, in addition to the tens of thousands of legitimate crossings every day. Pilot programs in several key locations, such as a crossing near Tijuana, have been established to test facial-scan and voice-scan systems to expedite the processing of frequent travelers.

The Dutch Immigration and Naturalization Department has implemented a system utilizing a smart card and locally stored finger-scan template. Established in 1993, the system is used by those seeking asylum in the Netherlands for identification at regular check-ins at automatic kiosks.

Employee Access

Several airports use biometric physical access devices to control employee access to secure areas. Examples include O'Hare Airport in Chicago, which uses a finger-scan system to control employee access to cargo, and the San Francisco airport and Hawaii airport, both of which use hand-scan to control employee access to secure areas. In July 2000, North Carolina's Charlotte/Douglas International Airport deployed an iris-scan system for access control, replacing traditional technologies. A similar solution was deployed in Frankfurt, Germany, in early 2001. Both projects are described as highly successful, with no false matches having occurred.

Passports

A largely untapped application of biometrics is passport issuance and processing, areas that could benefit in many ways from the inclusion of a biometric template on a barcode or smart chip. Guatemala, for example, has established a passport system using Printrak's AFIS. The system is used to ensure that only one passport is issued to each citizen and also allows users to have lost or stolen passports reissued at several U.S. airports using finger-scan verification.

Conclusion

September 11th has prompted a complete rethinking of the role of biometrics in travel and immigration. Surveillance, access control, passenger identification, crew identification, and "trusted travel" programs are all under investigation at U.S. airports. The U.S. government has also moved to incorporate biometrics in travel documents, increasing its ability to track visitors to the U.S. Biometrics are seen as one of few technologies that can help secure facilities and aircraft.

Additional Biometric Verticals

Though less central to the biometric industry than the five key verticals just described, the gaming and education verticals are well suited to biometric authentication and have seen some notable biometric deployments. In gaming applications, biometrics have found a steady foothold in many major casinos. Security and fraud are clearly major concerns, with many casinos going to great lengths to identify and expel cheaters and card-counting teams and to control access to controlled areas. In education, the need to authenticate students and guardians has led to some interesting deployments of biometric technology.

Gaming: contributory database surveillance. Facial-scan is the preferred technology for locating known cheats because of its ability to perform 1:N matching and to operate covertly. Three vendors—Biometrica, Imagis, and Griffin Investigations—currently supply facial-scan biometric systems to casinos. Although the facial-scan technology varies among vendors, the solutions provided by different vendors all function in a similar fashion, searching faces at gaming tables against a database of prohibited players. The technology is deployed at the majority of leading casinos in Las Vegas and in various other establishments.

Education: remote student authentication. Biometrics are being considered as a solution in remote learning environments, where students attend class or take tests over the Internet. Biometrics can ensure that the student being tested is the authorized individual. ContinuedEd.com licensed keystroke-scan technology to verify the identity of attendees of its online safe driving classes. It is likely that biometrics will be widely implemented as a solution in collegiate environments; though not an unbeatable system, it will act as a deterrent to cheating.

Education: student services. The British Columbia Institute of Technology issued smart cards with facial-scan technology for use as student ID cards in June 2000. The cards enable student access to locations, equipment, and facilities around the campus. Also, the cards can be used to store money for use at vending machines, copy centers, and other campus services. At the University of Georgia, hand-scan devices have been in use for over 10 years, implemented for students participating in prepaid meal plans; ID cards and other nonbiometric verification technologies are more prone to abuse or fraud. Welsh Valley Middle School in Pennsylvania installed a finger-scan system to track students' cafeteria purchases. Instead of paying for food in cash, students may use the finger-scan peripherals. At the end of each month, their parents receive a bill for the month's food purchases, or the food bill is sent to a free-food program for reconciliation. The deployment was prompted by legislation that forbids identification of students who participate in federal free-lunch programs. With students paying for lunch via a finger-scan system installed in cafeteria checkout lines, there is no way to determine whether their parents or government grants are paying for their meal.

Education: security. In the late 1990s in the United States, news of shootings and other violent incidents in schools received media attention; schools and parents are, as a result, firmly focused on protecting the safety of the students. Controlling school access with biometric devices increases the safety and perceived safety of schools and students. Charter Schools of Fort Lauderdale, Florida, deployed a finger-scan system for students and staff, used to verify individuals each day as they enter the school. After the day of classes begins, the doors are locked and can be opened only by an enrolled student or staff member. Harriet Beecher Stowe Elementary School in Chicago has implemented a similar system for 137 staff members, protecting the entrances to two school buildings.

Education: guardian authentication. Another aspect of protecting children comes into play in day care centers, nurseries, and elementary schools. A large percentage of kidnappings are committed by adults who the children either know or are closely related to, but who are not legal guardians. In locations where parents or guardians must come to pick up children, biometric systems have been deployed to verify their identity, preventing potential situations in which a child is removed from the premises by someone who is not a legal guardian.

Conclusion

Certain verticals are more likely than others to be areas of adoption for biometric solutions, driven by legislative initiatives, desire for cost savings, and the need for increased security. However, while biometric vendors have positioned their products as solutions particularly well-equipped for a specific vertical market, the core processes involved in biometric applications do not always differ substantially from vertical to vertical.

PART FOUR

Privacy and Standards In Biometric System Design

Two areas critical to the development of the biometric industry are privacy and standards development. Biometric technologies, because they represent characteristics unique to individuals, are frequently thought of as a privacy-invasive technology. Biometrics *can* be deployed in ways harmful to personal and information privacy. However, there are a number of inherently privacy-protective features of biometric technology which limit the likelihood of privacy-invasive usage. In addition, deployers can make system design decisions that turn a potentially privacy-invasive system into one that ensures the privacy of individuals' biometric information. Addressing the real and perceived privacy risks of biometric deployments is a critical task for any institution considering biometric usage.

Standards development is an essential sign of maturity for any emerging industry. Industry-approved guidelines for security, data formats, and application development reduce the risk of the unknown faced by institutions considering biometric deployments. The acceptance of biometric standards is essential to the technology's growth in sectors such as government, financial services, and e-commerce.

The chapters in this section examine the relationship between biometrics and privacy from a deployer's perspective, detailing the factors which determine whether a system is privacy-invasive, privacy-neutral, privacy-sympathetic, or privacy protective. In addition, the role that key standards play in the biometric industry is examined.

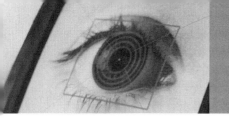

CHAPTER 15

Assessing the Privacy Risks of Biometrics

Privacy (n)—freedom from unauthorized intrusion; state of being let alone and able to keep certain esp. personal matters to oneself.

Merriam Webster's Dictionary of Law, 1996

Because biometric technologies are based on measurements of physiological or behavioral characteristics of the human body, discussion of biometrics frequently centers on privacy concerns. The increasing use of biometric technology raises questions about the technology's impact on privacy in the public sector, in the workplace, and at home.

➢ What are the major privacy concerns associated with biometric usage?

➢ What types of biometric deployments require stronger protections against privacy invasiveness?

➢ What biometric technologies are more susceptible to privacy-invasive usage?

➢ What types of protections are necessary to ensure that biometrics are not used in a privacy-invasive fashion?

Addressing these questions is necessary to ensure that biometric system design and deployment does not undermine or threaten personal or informational privacy. It is possible to be both a privacy advocate and a biometrics advocate: With appropriate protections and controls, biometrics can be implemented in a fashion consistent with privacy principles.

Privacy is a difficult concept to define, because privacy requirements change along with developments in technology, the willingness of the public to forgo a degree of privacy for the perception of increased security or safety, and the

237

degree of trust individuals have for public and private institutions. Deployers do not need an in-depth understanding of privacy to implement privacy-sympathetic biometric systems, but awareness of key privacy concepts—linking personal information, anonymity, and function creep—is a necessity.

Biometric Deployments on a Privacy Continuum

Before addressing the specific privacy concerns associated with biometric deployments, the different ways in which biometrics and privacy can be related must be defined. Biometric deployments can be privacy invasive, privacy neutral, privacy sympathetic, or privacy protective (see Figure 15.1).

At worst, biometric deployments can be *privacy invasive.* Biometric technology can be used without individual knowledge or consent to link personal information from a variety of sources, creating individual profiles. If unregulated, these profiles may be used for a variety of privacy-invasive purposes, such as tracking movement, associating unrelated data, or eliminating a person's ability to operate as an anonymous individual. Privacy invasiveness can be actual or potential—systems *capable* of being used in a privacy-invasive fashion are often perceived as privacy invasive.

Many biometric deployments are *privacy neutral*, lacking any special precautions or design elements to ensure privacy but also incapable of being used in a privacy-invasive fashion. In particular, certain biometric technologies are frequently deployed as privacy neutral: They cannot be used to protect individual information and cannot be used to undermine privacy.

Privacy-sympathetic deployments incorporate special design elements and controls to ensure that the biometric data cannot be used in a privacy-invasive fashion. A biometric deployment that encrypts biometric data, requires multiple administrators to access biometric databases, and stores biometric data independently of personal data is taking key steps toward being privacy sympathetic. Though such deployments may not be perfectly protective of privacy, they entail enough active protections and controls to make misuse of biometric data very unlikely.

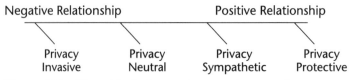

Figure 15.1 Biometric deployments on a privacy continuum.

Privacy-protective deployments are those that utilize biometric authentication to protect other personal information that may otherwise be susceptible to compromise. Giving individuals the option of accessing bank accounts, medical data, or other personal files through biometric authentication, as opposed to password or card-based authentication, is a typical privacy-protective deployment; these accounts are generally less likely to be compromised. Similarly, requiring that employees use biometrics to access sensitive files is a privacy-protective implementation, as audit trails are created that deter employees from accessing other individuals' data.

To illustrate how biometrics can be implemented in both privacy-invasive and privacy-protective fashions, consider how the Health Insurance Portability and Accountability Act (HIPAA) refers to biometrics. On the one hand, biometrics are cited as a potential identifier, such as a social security number, which must be disassociated from personal medical information before such information is shared for research purposes, lest biometrics be used to produce a unique identifier. On the other hand, biometrics are cited as one of several methods of entity authentication, which healthcare facilities can implement to protect computers and networks from employee abuse. It is the application of biometrics, not the technology itself, that defines its relation to privacy.

Privacy Concerns Associated with Biometric Deployments

There are two types of privacy concerns commonly expressed regarding biometric systems and technology. The first, informational privacy, relates to the unauthorized collection, storage, and usage of biometric information. The second, personal privacy, relates to an inherent discomfort individuals may feel when encountering biometric technology. While deployers must be prepared to address both types of concerns, informational privacy is viewed as the more critical issue in biometric implementations.

Informational Privacy

Today's computing technologies allow public- and private-sector institutions to gather, store, and compare a broad range of information about individuals. Using identifiers such as name, address, and social security number, institutions can search databases for information on a given individual. This information may relate to employment history, financial history, medical history, purchasing habits, and personal interests. Though frequently beneficial, such as in the case of emergency medical information being shared between hospitals, this ability to compile and share information about individuals can be very easily abused. The existence of searchable databases allows institutions to create profiles of individuals with information collected from a variety of sources. These profiles may enable employers to make hiring decisions based on private medical information or enable insurers to set higher rates for individuals who have conducted research on certain medical conditions.

Because of the risks involved in unauthorized linking of personal data, *informational privacy*—the right of an individual to exercise consent and control over the collection, storage, usage, and disclosure of data relating to him- or herself—is a major concern of privacy advocates.

The usage of biometric data, in and of itself, is not the problem. It is the potential linkage, aggregation, and misuse of personal information associated with biometric data, and stored in databases referenced by biometric data, that have a direct privacy impact. A fingerprint is meaningless until associated with an individual. The *potential* that exists in this equation is critical: From a privacy perspective, it is not enough that personal information is not linked and aggregated inappropriately. If it is theoretically possible that data might be linked, the data poses a privacy risk. Privacy-sympathetic biometric system design should, ideally, make linkage and association of personal data impossible under any circumstances, even if new laws are written that allow such linkage. Whereas today's lawmakers are relatively aware of the importance of privacy, in the future there may be increased temptation to use biometric systems for purposes beyond their original intent. Intelligent biometric system design limits the possibility that public- or private-sector institutions will have access to privacy-invasive tools.

Biometrics are seen as especially threatening to informational privacy, since the technology is often positioned by biometric vendors as the identifier you can't lose. This unchangeable identifier, in theory, could be used to track information about an individual across databases, from workplace to private life. Among the main risks that public and private institutions deploying biometrics could pose to informational privacy are unauthorized use, unauthorized collection, unnecessary collection, and unauthorized disclosure.

Unauthorized Use of Biometric Information

Unauthorized uses of the biometric technology are seen to represent the greatest risk biometrics pose to privacy. It is not normally the intended uses of biometrics that are seen as problematic—for example, a biometric system that ensures that an individual can receive only one driver's license would likely not meet with strong objections. However, the existence of a government database with facial-scan or finger-scan data may be tempting to law enforcement agents or to private-sector companies searching for personal data. It is this risk of unauthorized use of biometric data beyond the original intent for which it was collected—especially forensic usage and unique identifier usage—that defines many informational privacy fears.

Conducting criminal forensic searches on private-sector or nonforensic biometric databases (such as a driver's license database) would be highly problematic from a privacy perspective, as it would represent a substantial expansion of the government's ability to conduct searches. Given that use of fingerprints is the primary means of forensic identification, it is natural that privacy advocates question programs in which individuals provide finger-scan data in order to receive public benefits. The fear is that information provided for public- or private-sector usage will facilitate police searches, whether through digital fingerprint images or through latent prints taken from surfaces. Every biometric database becomes a potential database of criminal records, representing a significant increase in the government's investigative powers.

Using biometric data as a unique identifier is the second major unauthorized usage fear. A unique identifier is a fixed number or value, such as a social security number, associated with a specific person and capable of being used to locate information in a database. Unique identifiers are a risk in a world where nearly every fact about an individual—purchases, health information, financial information—is stored in some type of database. Unique identifiers can be used by malicious parties to monitor, link, and track a person's daily activities across disparate databases and information. When considering the various environments where one might provide biometric information in the public or private sector—banking, medical, public service, retail, and employment—the prospect of information linkage and collection is problematic. Because many biometric technologies are based on unique physiological or behavioral characteristics, the fear is that biometric technology can thereby serve as a unique identifier (see sidebar, *Are Biometrics Unique Identifiers?* on page 262).

Unauthorized Collection of Biometric Information

Although only certain biometric technologies are even capable of collecting biometric information without the subject's knowledge, the increased deployment of certain types of biometric technologies does bring with it the possibility that institutions could gather and process biometric information without consent. Facial-scan technology, voice-scan technology, signature-scan technology, and keystroke-scan technology, because they utilize standard devices (cameras, telephones, etc.) to acquire biometric information, could be used in this fashion. To this point, only facial-scan has been used to collect biometric information without user authorization. In some implementations—surveillance, for example—authorization to collect biometric information can be implicit. Signage indicating that a biometric system is in use in a certain area may be adequate to authorize any subsequent collection of biometric information.

Unnecessary Collection of Biometric Information

Biometrics are normally deployed in order to address a specific identity verification problem: controlling physical access to specific locations, controlling logical access to specific data, or ensuring that an individual does not enroll multiple times in a single-identity system. From a privacy perspective, deploying biometrics in environments in which they provide ill-defined or nominal benefits can undermine informational privacy. Unnecessary collection of biometric information contradicts a basic privacy principle that personal information should be collected only for specific reasons under specific conditions. Unnecessary collection of biometric information would most likely also facilitate unauthorized use.

Unauthorized Disclosure

Unauthorized disclosure—an institution's sharing of biometric information without a user's explicit consent—violates a fundamental privacy principle: that an individual has the right to exercise control over his or her own personal information. If biometric information is shared without an individual's authorization, the potential uses to which the data will be put, the information with which it is linked, and the security measures used to protect his or her biometric information are all unknown.

Function Creep

All of these informational privacy fears can be categorized as different types of *function creep*: the expanded use of a technology, system, or deployment for purposes beyond those originally intended. The expanded use of the social security number illustrates the danger of function creep, as many institutions collect social security numbers for authentication purposes well beyond those originally intended. Information-gathering services have been able to use this identifier to locate and link information across databases. Although linking this type of data is a necessary and frequently beneficial component of modern life, the potential dangers of function creep from a biometric perspective are more nefarious. At its most extreme, lack of regard for informational privacy could provide institutions with the ability to track individual movement and behavior. Biometrics could then be used as a tool that institutions could use to oppress individuals.

> ## Is Biometric Data Personal Information?
>
> Within the biometric industry, there is debate as to whether biometric data is personal information. On the one hand, biometric templates are encoded, proprietary files, which cannot be used to re-create identifiable images. Templates can be used as barriers to and protectors of personal information such as medical data and, as such, are not personal information, but robust keys used to unlock personal information. On the other hand, biometric data is derived from an individual's behavioral or physiological characteristics, is used to verify or determine a person's identity, and is strongly associated with, in cases unique to, each individual. This makes biometric data, even in template form, personal data.
>
> Whether or not one views biometric data as personal information, it is incontrovertible that those opposed to or unfamiliar with biometric technology will view it as such. It is difficult for biometric deployers or advocates to address user concerns about biometrics by suggesting that biometric data capable of identifying an individual from a database of thousands or millions of users is not personal information. Those looking to implement biometrics must design and deploy systems on the assumption that biometric data is personal data and should be protected accordingly.
>
> The fact that biometric data should be treated as personal information is not an objection to its use. There are countless situations in which the collection, storage, and usage of biometric data by individuals, employers, or government agencies are beneficial. Biometric data is a type of personal information that, if used properly, can protect other, more sensitive, personal information. This capability of biometrics to protect sensitive personal information, such as records pertaining to health, employment, or finances, is the primary basis of biometrics being positioned as a privacy-enhancing technology. However, because biometric data is sensitive and there are situations in which biometric systems could be misused, protections tantamount to the deployment-specific risks are necessary at all possible stages of the data's life cycle.

These informational privacy risks are real, but there are a number of protections biometric deployers can implement to significantly reduce the potential privacy impact, as seen subsequently.

Personal Privacy

In addition to the concerns categorized as informational privacy, biometrics are also objected to on the grounds of personal privacy. There is a percentage of the population for whom the use of biometric technology is inherently offensive, invasive, or disturbing. While some personal privacy fears may be derived from inchoate informational privacy concerns, this reaction to biometrics is often attributable to cultural, religious, or personal beliefs. Other individuals may feel that the implementation of a biometric system in the workplace, for example, is an expression of mistrust of employees and find the implicit mistrust insulting.

Many new technologies are met with strong objections upon their introduction. ATMs, for example, were looked at with suspicion for years. Smart cards have long been viewed as an invasive technology in the United States and Canada; meanwhile, billions of smart cards have been issued across Europe, Asia, Africa, and South America. Although millions of individuals are enrolled in biometric systems of varying complexity and scope around the world, the technology is still perceived as futuristic and in some ways threatening.

For deployers, it is more difficult to countervail objections based on personal privacy than those based on informational privacy. Whereas objections based on informational privacy can be mitigated by describing how the system works and defining policies that protect biometric information, objections based on personal privacy are exactly that—personal. Methods of effectively addressing objections based on personal privacy vary whether the technology is being implemented to customers, employees, or citizens, and whether systems are mandatory or optional. Mandatory systems are more likely to be met with personal privacy objections, as individuals may feel coerced into enrollment and take a dim view of the technology. In this case, public- and private-sector deployers can make clear why biometrics are being deployed, where they have been deployed in similar situations, and how the systems work.

Over time, acclimation to biometric technology may reduce objections based on personal privacy. Individuals with experience in using a biometric system have consistently approved of the technology more strongly than those who have not used biometrics. Lack of familiarity with biometric technology seems to breed contempt; fortunately for the biometric industry, testing shows that once an individual has used a biometric, he or she is less likely to object to using it on an ongoing basis.

Privacy-Sympathetic Qualities of Biometric Technology

Many concerns regarding biometrics and privacy are well grounded, while others are based on fundamental misconceptions about the technology's operation. There are a handful of facts about biometric technology and the biometric industry that reduce the likelihood that biometrics can be used in a privacy-invasive fashion.

Most biometric systems (with the primary exception of forensic systems) do not utilize or store raw biometric data such as fingerprint images or facial images, but instead store biometric templates, significantly reducing privacy risks. Were every biometric system to store raw images, tracking data across databases would be much simpler. An individual's right index fingerprint, if stored for usage in home, work, and government-oriented systems, could easily be used to associate this information. Templates, by comparison, vary from placement to placement during both enrollment and verification, such that tracking becomes extremely difficult. These templates cannot be used to re-create the original images, but have enough information about the original image to provide a high confidence level in subsequent matches. As opposed to performing binary searches for identical strings of data, an institution would need to use matching algorithms to determine whether individual information in one database could be linked with information in another.

Imagine selecting the 10th, 20th, and 30th letters of a book, continuing to extract every 10th letter until one has a 40-letter template. It would be impossible to re-create the book from this sequence of letters, but very few books would generate the same template. Furthermore, imagine that the book—or biometric sample—changes very slightly over time, such that some words are misspelled or missing. Finally, imagine that one cannot determine whether two 40-letter strings are from the same book by simply comparing the letters from beginning to end, but that a specific algorithm must be used to locate similar patterns within the 40-letter string. This is the nature of biometric template generation and comparison, and should help explain why template-based biometric systems are highly resistant to privacy-related abuses.

The range of biometric technologies used for different types of applications limits the likelihood that a universal biometric identifier will be implemented. It is effectively impossible to biometrically link data from an individual who uses finger-scan to access networks, hand-scan to enter secure areas, and voice-scan to access telephone accounts.

The ability to enroll various biometric samples, especially in behavioral biometric systems such as voice-scan and signature-scan, also limits the likelihood of biometric data being used in a privacy-invasive fashion. A user can enroll different biometric samples in behavioral biometric systems. For example, one could utilize a full signature for banking functions, but enroll a first initial-last name for authorizing prescriptions. The number of possible enrollments in a behavioral biometric system is unlimited.

Finally, the fact that the biometric industry comprises dozens of companies with proprietary, noninteroperable technology reduces the likelihood of biometric data being tracked across databases. Different finger-scan systems are well suited for physical access, PC access, and smart card-based identification; several companies compete in each of these markets, resulting in dozens of types of biometric templates, which can only be compared to templates generated through the same technology.

Defining Application-Specific Privacy Risks: The BioPrivacy Impact Framework

One of the biggest oversights in the discussion of biometrics and privacy has been the failure to discuss the privacy impact of specific types of biometric deployments. Certain types of biometric deployments are more prone than others to lead to privacy-invasive uses, while other types of deployments have little or no bearing on privacy. Biometrics, in and of themselves, are neither a protector nor an enemy of privacy; instead, the type of deployment determines the relation between biometrics and privacy. However, the issue of biometrics and privacy has been addressed as if all biometric deployments were identical. The potential privacy impact of a voluntary biometric system, with templates stored only on a user-possessed smart card, differs dramatically from that of a mandatory system with central storage.

In order to help deployers define the potential privacy impact of a biometric deployment and take appropriate precautions to ensure that deployments are privacy sympathetic, the BioPrivacy Impact Framework was developed in mid-2001. The BioPrivacy Impact Framework, when used in conjunction with BioPrivacy Technology Risk Ratings and BioPrivacy Best Practices, allows deployers to implement the proper controls on biometric data and system design. Assessing a biometric deployment through the BioPrivacy Impact Framework illustrates the areas where greater risks are involved, such that appropriate precautions and protections can be enabled (see Table 15.1). Without the Impact Framework, system designers may feel compelled to implement the same privacy protections for a card-based, hand-scan time and attendance system as for a public-sector, 1:N facial-scan system; to do so would waste time and effort, and would likely lead to less privacy-sympathetic systems.

Overt versus Covert

Deployments in which users are aware that biometric data is being collected for storage and/or comparison, and in which acquisition devices are in plain view, are less privacy invasive than surreptitious deployments. User awareness and consent are key principles of privacy-sympathetic deployment, and it is difficult to consent to covert systems. Notices and signs indicating the use of covert biometric systems in public spaces, for example, can allow individuals who do not consent to such usage to avoid those areas. Systems that are completely covert—wherein notice or signage would render the system ineffective—may be necessary, but should be limited to situations in which there is a highly compelling public interest. The issue of overt versus covert deployments applies primarily to facial-scan technology, which can utilize existing surveillance camera technology to identify individuals in a covert fashion.

Table 15.1 BioPrivacy Impact Framework

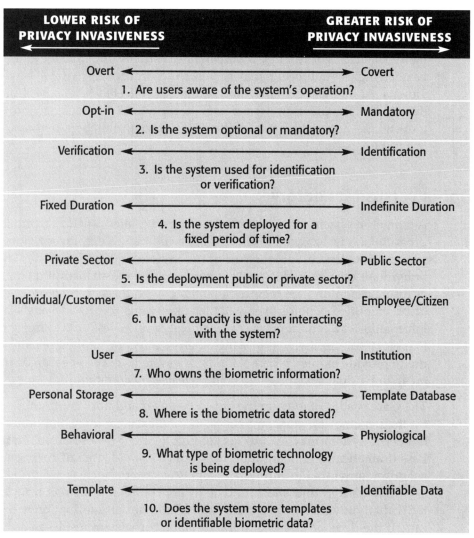

When designing and deploying covert systems, institutions must determine whether their primary objective is to deter criminal behavior or to detect criminals. The deterrent effect of biometric systems will clearly be enhanced with notices and signage, but the ability to detect criminals will likely be reduced. Criminals will either avoid the area under surveillance or take measures such as altering their appearance to avoid detection. Without signage or notices, the criminals presumably will be unaware of covert biometric operations and will not avoid areas under surveillance. While balancing these considerations, deployers must determine the potential privacy impact of not providing notice and signage.

If a covert system *is* deployed without signage, special protections must be enacted to ensure that the system cannot permanently store biometric information collected from those individuals who do not match watch lists. If biometric databases of passersby were maintained, the potential for abuse would be quite high, as authorities may be able to arbitrarily search databases for individuals. At its most extreme, storing images collected from covert biometric deployments could represent a tool to monitor and track an individual's movement—acceptable in an airport or military installation, perhaps, but not in a public space.

Opt-in versus Mandatory

A biometric system in which enrollment is mandated, such as a public-sector program or one designed to enroll all of a company's employees, runs greater privacy risks than a voluntary or opt-in system. In mandatory deployments, individuals are required to provide biometric data to an institution for storage or future usage, and are required to interact with the system in order to verify or be identified. Choice over whether one wants to provide one's personal information is a central privacy principle. Users are less able to freely consent to a biometric deployment in which there is a real or implied penalty for failure to provide biometric data. Mandatory systems may appear to be imposed on an individual as opposed to being selected by an individual.

There are shades of gray on the opt-in versus mandatory continuum, defined by the penalties for noncompliance. At one extreme, noncompliance with a biometric system may lead to loss of employment or loss of public benefits. Less dramatically, an institution may apply some degree of coercion to persuade an individual to enroll in a biometric system, such as requiring a more extensive background check in case of noncompliance or requiring that the individual provide a greater-than-normal amount of information to authenticate. If the decision not to enroll in a biometric system results in any sort of punitive measure, it is not truly voluntary, and the concept of consensual enrollment is undermined.

In the real world, many biometric deployments are effective from a security or fraud prevention perspective only if they are mandatory for all applicable employees or citizens. A biometric system designed to locate duplicate enrollments in a database (such as a public benefits program) would be useless if it were optional—users committing fraud would simply not enroll. The security of employee-facing systems, such as PC access and physical access, is undermined unless such systems are mandatory and universal. If a percentage of users decide to use badges or passwords instead of a biometric system, the result is a significant gap in the system's ability to secure data or a physical location. The system can be compromised through those accounts that utilize weaker authentication methods. Employee-facing applications in which convenience, not security, is the primary driver are much more likely to be voluntary, as the system is not undermined by having a mix of biometric and nonbiometric authentication. Customer-facing applications—point of sale, e-commerce, and telephony, for example—are also much more likely to be voluntary. A large percentage of customers may not have access to a biometric device or simply may not want to enroll in biometric systems. Deployments entailing voluntary customer enrollment are less likely to pose privacy risks, although appropriate protections must still be in place.

While both mandatory and optional systems must be designed to resist misuse of personal data, each type of system is susceptible to particular types of misuse. Data from users who voluntarily enroll in systems may be viewed as more shareable. Since users opted to utilize a biometric for a given system, an institution may conclude that users wouldn't mind such data being used for broader purposes. The key protection required in this instance is against unauthorized use. In a mandatory system, it is more likely that open biometric searches may be conducted for employees or citizens whose data was required for a specific program. Protections against both unauthorized use and disclosure are among those required in mandatory implementation.

Verification versus Identification

A system capable of performing $1:N$ identification—able to search databases for a specific individual using only biometric information—is more susceptible to privacy-related abuse than a system only capable of 1:1 matching. From a privacy perspective, the ability to search a database using only biometric information can lead to increased ability to associate an individual with certain activities or behaviors. For example, a high-quality finger-scan database used in a public-sector benefits program could be searched to identify criminals or to locate an individual with a specific medical condition, without the individual's consent. When biometric information in $1:N$ databases can be associated with specific user data, such as name, address, medical history, employment, and citizenship status, the potential risks of abuse grow.

In addition to increased risk of data aggregation, the ability to identify individuals reduces an individual's ability to transact with a degree of anonymity. There are environments in which a user may want to maintain a home identity to access personal information on the Internet, and a work identity to conduct business transactions, access networks, and so on. The ability to perform identification using only biometric data increases the likelihood that these identities can be linked.

Not all biometric technologies are capable of operating in 1:N mode. Hand geometry, for example, can verify identity, but cannot identify individuals based on their biometric characteristics. In order to operate in 1:N mode, a biometric must have a sufficiently high number of distinctive and stable characteristics. Biometric systems capable of being used in identification mode are always centralized; that is, biometric data is stored in a database as opposed to being stored on a user's local PC or smart card. This centralization is an additional privacy risk that must be countered by privacy-sympathetic system design. To counter the risks involved in identification systems, controls must be implemented on who has access to the system, on the conditions under which searches can be conducted, and on the types of results that can be generated. In this instance, security policies enable system usage consistent with privacy expectations.

Fixed Duration versus Indefinite Duration

In deployments where such an option exists, the use of biometrics for a fixed duration is less likely to have a negative impact on privacy than one deployed indefinitely. This applies in particular to public surveillance deployments, which are more likely to bear a questionable relation to privacy than other biometric deployments. When deployed for an indefinite duration, the risk of scope creep increases. Biometric surveillance may then be viewed as a commonplace as opposed to an exceptional event. The privacy risk is that a great deal of information will be collected over time and may be used to track individual movement.

Many biometric deployments, such as securing network login or PC login, are meaningful only when implemented permanently. In this case, special control must be taken to ensure that data cannot be used beyond the purposes for which it was originally collected.

Public Sector versus Private Sector

Although both public-sector and private-sector deployments bear their own privacy risks, the worst-case scenarios are found in public-sector deployments due to the possibility of state or government abuse of personal data. Government monopolies on law enforcement mean that the linking of personal information through biometrics can result in incarceration or imposition of physical force. The Big Brother fear frequently cited as an objection to large-scale biometric implementations is based on government collection and misuse of biometric data for purposes of tracking, monitoring, and controlling behavior. While such a fear may seem far-fetched in many Western countries where systemic controls on government power are in place, the potential for misuse in countries with histories of totalitarianism may be greater. A government body in possession of a comprehensive biometric database on all of its citizens could use the technology as a tool of oppression, even if the data were collected for innocuous purposes. Clearly, controls on certain types of public-sector government systems—in terms of how personal data can be used and who can access the system—are essential.

However, private-sector deployments are not without potential privacy impact. While private-sector biometric databases are less likely to be abused by law enforcement, they are more likely to be shared for marketing or commercial purposes without individual consent. A biometric enrollment is a piece of personal information that may have substantial value, especially if it can be associated with purchasing habits or other information. Different types of controls, especially in terms of consumers' control over their biometric information, are necessary to ensure that private-sector deployments are consistent with privacy expectations.

Citizen, Employee, Traveler, Student, Customer, Individual

An individual's roles vary according to the people and institutions with which he or she interacts. Individuals are citizens or residents in their dealings with the government, employees in their dealings with employers, customers in many commercial transactions, and simply anonymous individuals in many environments. The role an individual occupies when his or her identity is authenticated bears directly on potential privacy impact, as certain roles are more likely to be associated with personal information and with potential misuse than others. In addition, some types of institutions are in a position where they can dictate how users authenticate and can implement penalties for noncompliance.

Citizen, employee, traveler, student, customer, and *individual* are the primary capacities in which an individual may interact with an institution. The relationship between the individual and the authenticating entity varies with each category; here, they are listed in descending order of the potential penalties for noncompliance with the biometric system. For example, the penalties that might be imposed on a citizen for noncompliance, such as loss of benefits or even criminal charges, are greater than penalties an employee might face for noncompliance, such as reprimand or termination. This results in the biometric being associated with compulsion and reduced individual choice. Because of the potential sanctions involved, an individual is less likely to have choice over whether to use a biometric system as a citizen or as an employee. Reduced choice over one's information is at odds with privacy principles and must be countered with special protections to ensure that such data cannot be misused by design or default.

To address other individual roles, an individual acting in the capacity of a traveler faces less severe penalties for noncompliance than when acting as a citizen or employee. The potential sanction is the loss of the ability to travel by a selected means—the person may not be able to board a plane but could find another method of transport. There is a reduced expectation of privacy in a travel application, as there is a compelling interest to authenticate passenger identities, especially in air travel. Students, too, bear a special relationship to the authenticating entity such as a school or university. The entity has some sanction over the student's behavior, and the potential range of penalties for noncompliance with school regulations can range widely. Depending on their age, students may or may not be able to freely consent to a biometric system in the legal sense. Because of this, protections are required to ensure that such systems are implemented only when necessary and are, if possible, noncompulsory.

When an individual is biometrically authenticated as a consumer, the challenges of maintaining individual privacy change. Consumer-facing biometric usage is almost always optional, as consumers are more likely to dictate the conditions of the transaction than merchants. If consumers want increased security and convenience, biometric technology may be available, but generally as an option. Any merchant making biometric usage mandatory risks driving some portion of customers to competitors. From a privacy perspective, the risks have more to do with control of personal data than with compulsory enrollment.

Individuals authenticating simply as individuals—logging into anonymous email accounts, selling items to other individuals through auctions—are not authenticating to a larger entity, but instead are assuming a persona to exchange ideas or goods. The privacy risk in this case has less to do with inability to freely choose how and when one's personal information is collected than with the unauthorized sharing of such data. An anonymous biometric account, if linked with another biometric account, is no longer anonymous.

Although privacy rights are fundamental regardless of the institution with which the person is interacting, they are not identical in all environments. Reasonable expectations of privacy are dependent on the capacity in which a person is interacting with another person or institution. To counteract this, biometric systems deployed in each of these environments must be designed and controlled according to the potential risks involved for the user population. Above all, to ensure that a person can maintain his or her separate identities, data residing in separate biometric systems must not be linked or amalgamated without the explicit, informed permission of the individual.

User Ownership versus Institutional Ownership of Biometric Data

Deployments in which the user maintains ownership over his or her biometric information are more likely to be privacy sympathetic than those in which the public or private institution owns the data. User control over collection, usage, disclosure, and disposal of biometric information is not possible in every deployment, especially in entitlement programs or other public-sector uses. However, when individuals maintain ownership of biometric data and can choose how it is managed and whether to continue to utilize the biometric system, the potential privacy risks are significantly reduced.

An individual may own his or her biometric data without having direct possession of the biometric templates. Ownership implies the right to determine how and when data is used; biometric data may be stored in a company database but still be owned by the individual in an opt-in scenario. Institutional ownership of biometric data increases the risk that data will be shared or disclosed without individual consent.

Personal Storage versus Storage in Template Database

A biometric system that stores information in a central database—whether central to an employer, a merchant, or a government organization—is more capable of being abused than one in which biometric information is stored on a user's PC or on a smart card. Most scenarios involving misuse of biometric data entail a database that has been compromised or searched in an unauthorized fashion. These searches may be conducted by law enforcement officials or by internal agents. The risk of open or arbitrary matches on a series of biometric templates is eliminated when biometric data is not stored centrally.

Decentralized storage does bear its own smaller-scale privacy risks. When biometric data is stored locally, there may be increased risk that a single template is compromised, because the type of protections available to large-scale systems may not be present in home or workstation PCs. In addition, biometric data, when stored on a local PC, may be the personal information that an individual has stored locally. When biometric data is stored centrally, it is almost always segregated from any personal data.

Many systems require that biometric data be stored centrally for security purposes. If the template is under institutional control, its integrity cannot be undermined, and more secure auditing can be enabled. Institutions that store biometrics for use in customer-, employee-, and citizen-facing systems are generally well versed in protecting sensitive data. The emergence of trusted third parties—institutions whose core competencies include storing and handling sensitive data in a secure fashion—is meant to address the challenges of central biometric storage. Institutions that do not want to assume the risk of storing biometric data but still require that the template not be under the control of the user can turn to trusted third parties to assume the challenges of template management.

Behavioral versus Physiological Biometric Technology

Physiological biometrics, in general, are more likely to be used in a privacy-invasive fashion than behavioral biometrics. Physiological biometrics such as finger-scan, iris-scan, hand-scan, and facial-scan are based on relatively stable characteristics and are more likely to be traceable over time. These characteristics are much harder to mask or alter than behavioral biometrics and can be collected without user compliance. Behavioral biometrics such as voice-scan and signature-scan can be easily changed by altering a signature or using a new passphrase; authentication is contingent on users voluntarily providing a biometric sample. Because of the high degree of variability in behavioral biometric samples, these technologies cannot be used in $1:N$ applications, providing another privacy protection. All else being equal, deployments that use physiological biometrics (in particular finger-scan and facial-scan) require more robust protections than those based on physiological biometrics.

Template Storage versus Identifiable Data Storage

Most biometric systems store biometric templates as opposed to identifiable biometric samples. Many biometric privacy protections are predicated on template usage, as previously described. Templates cannot be used to generate identifiable biometric samples, and they require specific matching algorithms to compare data. Identifiable biometric data such as fingerprints or facial images can be compared without proprietary matching algorithms.

In many deployments, identifiable biometric data must be stored in order for the system to be operable. For example, law enforcement systems must store fingerprints along with finger-scan templates in order to resolve borderline matches, so that an expert can compare two fingerprints to manually determine whether there is a match. Driver's license systems must store images in order to print cards and to allow for visual inspection of applicants. Special protections such as data encryption should be enabled to ensure that identifiable data cannot be compromised.

Conclusion

The BioPrivacy Impact Framework provides a means of assessing the privacy risks involved in biometric deployments. Instead of treating all biometric deployments as bearing equivalent privacy risks, the Impact Framework defines the types of deployments most susceptible to privacy-invasive usage. Clearly, a private-sector biometric application in which the user retains ownership of his or her biometric information is much less likely to negatively impact user privacy than a covert public identification system; the precautions taken in each system must be proportional to the potential risks. Though there are many additional factors to assess when evaluating a deployment's privacy impact, such as the political climate and legal backdrop for biometric usage, the Impact Framework provides a starting point for intelligent assessment and implementation of biometric systems.

BioPrivacy Technology Risk Ratings

Just as each type of biometric deployment can have a different impact on privacy, each biometric technology bears a different relation to privacy. Some technologies have almost no privacy impact and would be difficult to use in any privacy-invasive fashion. Other technologies are much more likely to be associated with privacy-invasive usage, either because of their basic operations or due to extrinsic factors.

The BioPrivacy Technology Risk Ratings assess the privacy risks of leading biometric technologies, defining which technologies require more explicit system-design protections than others. The Technology Risk Ratings evaluate biometric technologies according to three categories presented in the Impact Framework: *Verification/Identification*, *Overt/Covert*, and *Behavioral/Physiological*.

In addition, the Technology Risk Ratings include a fourth category referred to as *Give/Grab*. Biometric systems can acquire biometric data in two ways: (1) when an individual *gives* biometric data at the time of his or her choosing after initiating an enrollment or verification sequence or (2) when a biometric system *grabs* biometric data without the user having initiated an enrollment or verification system. The difference may seem slight, but systems in which users are able to decide precisely when they provide biometric data are more consistent with privacy principles.

Each technology is given a Risk Rating of *Low*, *Medium*, or *High* in each of the four categories.

Verification/Identification. Technologies only capable of verification are rated lower; technologies capable of robust identification are rated higher.

Overt/Covert. Technologies requiring that individuals be aware of biometric system operation are rated lower; technologies capable of operating without user knowledge or consent are rated higher.

Behavioral/Physiological. Technologies based on variable behavioral characteristics are rated lower; technologies based on unchanging physiological characteristics are rated higher.

Give/Grab. Technologies in which the user gives biometric data are rated lower; technologies in which the system grabs user data without the user initiating a sequence are rated higher.

Low: Little if any privacy risks. The basic functionality of the technology ensures that there are few if any privacy issues.

Moderate: Limited privacy risks. The technology could be used in a privacy-invasive fashion, but the range of potential misuse is limited.

High: Substantial privacy risks. Without proper system design protections, the technology could be used in a privacy-invasive fashion.

As the Technology Risk Ratings table (Table 15.2) indicates, finger-scan and facial-scan pose high privacy risks; iris-scan, retina-scan, and voice-scan pose moderate privacy risks; and signature-scan, keystroke-scan, and hand-scan pose low privacy risks.

It would be a mistake to conclude from the Technology Risk Ratings that finger-scan and facial-scan deployments should be limited and that lower-risk technologies are preferable. In most situations, deployers do not have the luxury of choosing which biometric technology to implement. Factors such as accuracy, price, process flow, ease of integration into current systems, form factors, and existing authentication schemes normally drive deployers to implement a specific biometric technology. In many implementations, the only technologies that make sense are those that may be more susceptible to privacy-invasive usage. The lesson to draw is that certain technologies are more susceptible to privacy-invasive usage, and protections in place must address the technology-specific risks.

Table 15.2 Technology Risk Ratings

TECHNOLOGY	POSITIVE PRIVACY ASPECTS	NEGATIVE PRIVACY ASPECTS	BIOPRIVACY TECHNOLOGY RISK RATINGS
Finger-scan	• Large variety of vendors with different templates and algorithms • Can provide different fingers for different systems	• Use in forensic applications • Storage of images in public-sector applications • Strong identification capabilities	Verification/Identification: H Overt/Covert: M Behavioral/Physiological: H Give/Grab: M *Risk Rating: H*
Facial-scan	• Changes in hairstyle, facial hair, position, lighting reduce ability of technology to identify individuals	• Easily captured without consent or knowledge • Existing facial image databases can be used for comparison	Verification/Identification: H Overt/Covert: H Behavioral/Physiological: M Give/Grab: H *Risk Rating: H*
Iris-scan	• Requires high degree of user cooperation—difficult to acquire without consent • Requires proprietary acquisition device • Iris images not used in forensic applications	• Very strong identification capabilities • Development of technology may lead to covert acquisition capability • Only one type of iris template—no vendor heterogeneity	Verification/Identification: H Overt/Covert: L Behavioral/Physiological: H Give/Grab: M *Risk Rating: M*

Continues

Table 15.2 Technology Risk Ratings *(continued)*

TECHNOLOGY	POSITIVE PRIVACY ASPECTS	NEGATIVE PRIVACY ASPECTS	BIOPRIVACY TECHNOLOGY RISK RATINGS
Retina-scan	• Requires high degree of user cooperation—cannot be captured without consent • Requires proprietary acquisition device	• Very strong identification capabilities • Can indicate certain eye diseases	Verification/Identification: H Overt/Covert: L Behavioral/Physiological: H Give/Grab: L *Risk Rating: M*
Voice-scan	• Voice is text-dependent: user must speak a specific password to be verified • Not capable of identification usage	• Biometric data can be captured without consent or knowledge	Verification/Identification: L Overt/Covert: H Behavioral/Physiological: L Give/Grab: M *Risk Rating: M*
Signature-scan	• Signing is largely behavioral—can be modified at will	• Signature images can be used to commit fraud	Verification/Identification: L Overt/Covert: L Behavioral/Physiological: L Give/Grab: L *Risk Rating: L*
Keystroke-scan	• A highly behavioral characteristic—subject to significant day-to-day changes	• Can be captured without knowledge/consent	Verification/Identification: L Overt/Covert: M Behavioral/Physiological: L Give/Grab: L *Risk Rating: L*
Hand-scan	• Physiological biometric, but not capable of identification • Not a palm-scanner, but a measure of hand structure • Requires proprietary device	• None	Verification/Identification: L Overt/Covert: L Behavioral/Physiological: M Give/Grab: L *Risk Rating: L*

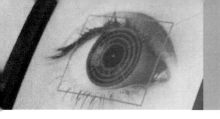

CHAPTER 16

Designing Privacy-Sympathetic Biometric Systems

Having discussed the privacy concerns most often associated with biometric systems, the inherent protections that biometric technologies can offer, and the types of deployments and technologies most often associated with privacy-invasive biometric usage, we can now investigate how to provide informational and personal privacy in biometric deployments. The tools we will use are International Biometric Group's BioPrivacy Best Practices.

IBG's BioPrivacy Best Practices define what steps institutions can take when deploying biometrics to ensure that biometric deployments do not intrude on individual privacy and are instead either privacy neutral or privacy sympathetic.

Very few deployers will be able to adhere to all BioPrivacy Best Practices during system design and implementation. Indeed, some deployers may be able to incorporate only a portion of these recommended protections and still operate an effective system. It is recommended that deployers implement as many Best Practices as possible without undermining the basic operations of the biometric system. When a proposed system is incapable of being designed and deployed in a fashion consistent with a specific Best Practice, reasons for noncompliance should be available for customer or client review.

Above all, deployer awareness of Best Practices is critical; some Best Practices may be incompatible with the basic process flow of a biometric system, but each decision to adhere or not to adhere to a specific Best Practice should be examined and justified. In this way, every aspect of a biometric system is examined from a privacy perspective, and privacy risks are minimized.

BioPrivacy Best Practices: Scope and Capabilities

The first challenge of privacy-sympathetic system design is to address the system's scope and capabilities: what the system is meant to do, and how it accomplishes this task.

Limit System Scope

Biometric deployments must not be expanded to perform broader verification or identification-related functions than originally intended. Any expansion or retraction in scope should be accompanied by full public disclosure, under the oversight of an independent body, allowing individuals to opt out of system usage if possible. A fundamental risk with any system of identification is that the system can be employed for purposes beyond those that were originally intended. From a privacy perspective, function creep must be disallowed, even if the purposes of the system expansion are seen as innocuous.

The scope of a biometric system can be limited by legislation, by internal or third-party oversight, and by the type of data collected. Systems can also be designed that preclude the artificial introduction of images or biometric data, requiring that a live fingerprint or facial image be presented in order for a decision to be rendered. However, because it is difficult to design a system that categorically cannot be used for purposes beyond its original intent, auditing, oversight, and transparency are essential. If a system is being misused, drawing attention to this misuse and enabling policies whereby system usage can be suspended are required. Scope limitation may be more difficult in countries with authoritarian governments, where frameworks to ensure public- and private-sector accountability may be lacking.

Do Not Use Biometrics as a Unique Identifier

The use of biometric information as a unique identifier should be extremely limited; in public- and private-sector systems, sufficient protections must be in place to ensure to the degree possible that biometric information cannot be used as a unique identifier. Unique identifiers facilitate the gathering and collection of personal information from various databases and can represent a significant threat to privacy. Though biometric templates are not ideal unique identifiers—a user's biometric verification differs every time he or she is authenticated—the enrollment template is normally a fixed value, used in all subsequent verifications. If a user's static enrollment template were shared between various agencies or companies to enable verification to a range of systems, it could be used as a highly effective unique identifier.

The unique identifier issue will become more problematic if a biometric technology is developed that generates the same template every time a user interacts with a system. This type of template could be used as an identifier across multiple databases and applications, and any single verification template could be linked with all of a user's verification templates.

Designing systems in which the enrollment cannot be exported to other systems and which do not store identifiable biometric data are two protections against the use of biometric information as a unique identifier. In addition, biometric data can be stored separately from other personal information such as account numbers, names, and addresses. This ensures that even if a piece of biometric data is compromised, an intruder will need to penetrate other databases to find any usable information. In situations where the use of a biometric as a unique identifier for circumscribed purposes is permitted, agencies that store and retrieve such data must be under the oversight of an independent agent capable of suspending system operation.

Limit Retention of Biometric Information

Biometric information must only be stored for the specific purpose of usage in a biometric system and should not be stored any longer than necessary. Biometric information and associated account data should be destroyed, deleted, or otherwise rendered useless when the system is no longer operational. However, data such as transactional logs can be kept for auditing purposes.

Different storage limitations apply to enrollment and verification data. While enrollment data, by definition, must be retained in order for the system to be operational, verification data need only be retained for as long as necessary to perform a match. Once a decision is rendered, there is no need to store the biometric verification attempt, and well-designed systems will dispose of verification data once a decision is rendered. In some situations, a hash of the verification template may be stored to prevent compromised templates from being used in replay attacks. What is essential is that biometric systems, whenever possible, delete a person's biometric data once he or she is no longer a transacting entity.

System design can accomplish some of this task by deleting biometric information when an associated account is deleted or updated. In addition, biometric systems can be designed to be incapable of storing verification templates. In many cases, only oversight and auditing can ensure that biometric data is not retained beyond its originally intended duration.

Are Biometrics Unique Identifiers?

A *unique identifier* is a fixed number or value associated exclusively with a specific person. In the United States, the social security number functions as a unique identifier, although social security numbers can be stolen or mistakenly issued to more than one individual. Because an individual's biometric data cannot be lost, stolen, changed, or forgotten, physiological biometrics such as fingerprints and iris patterns are often viewed as especially robust identifiers. However, while the biometric characteristics on which biometric systems are based may be unique, the templates that represent these characteristics within biometric systems cannot be used as unique identifiers. How is this possible?

Let's assume that the biometric system does not store raw biometric data such as fingerprints, iris images, or voice recordings (such data is normally stored only in law enforcement applications), but instead stores templates based on these characteristics. The templates generated from these physiological characteristics are unique *but nonrepeating.* As opposed to an identical string of data, biometric templates vary with each finger placement, iris acquisition, and voice recording: The same finger, placed over and over again, generates a different template with each placement. This is attributable to minute variations in presentation—pressure, distance, angle, pitch, skin condition, time of day—which lead to the extraction of slightly different features for each template. Without using a vendor's proprietary matching algorithm to process these templates, they cannot be identified as being from the same person. This prevents biometric identifiers from being traced across different databases. The idea that a unique biometric identifier can be tracked across every system in which an individual enrolls is inconsistent with the technology's basic operations.

If biometric templates were identical with each placement, then biometric thresholds would not be necessary—all matches and nonmatches would be 100 percent. Because of the variance in template generation, there is always the possibility that a comparison of two templates will result in a false nonmatch. Interestingly, the fact that biometric template generation is not perfect helps ensure that biometrics cannot be readily used to facilitate multidatabase tracking.

Since templates cannot be tracked by searching for repeating patterns, could an institution use biometric algorithms to compare a series of templates from different databases and find related data? Though not impossible, there are a number of impediments to such illicit tracking. Every biometric system would need to use the same vendor's technology; different vendors' algorithms and templates are not interchangeable or comparable in any fashion. Assuming that only one biometric company provided the core matching technology for all systems, the companies managing the databases—employers, retailers, trusted third-party providers—would need access to vendor source code in order to match templates. Even with the complicity of the biometric vendor and the companies responsible for storing and managing personal data, the ability to compare against large databases is limited by the distinctiveness of the biometric data. Comparisons against databases of several thousand users would begin to return multiple matches. Finally, this sort of tracking assumes that the individual has enrolled the same finger in each database. Some individuals purposely enroll their left index finger for home use and their right index finger for work, rendering impossible any type of tracking effort.

This entire discussion assumes that the biometric data in question is stable over time, which is not always the case—fingerprints can change slightly over time due to wear and tear. It also assumes that the biometric templates are stored centrally and not on a local PC or smart card. If the underlying biometric data changes or the templates are under the control of the user as opposed to the institution, tracking a unique biometric identifier becomes that much more unlikely.

For institutions deploying biometrics to employees or customers, making this distinction clear to users is essential. As users gain a better understanding of how biometric systems operate and what inherent protections are in place to ensure that biometric data cannot be readily tracked, they are more likely to respond positively to the system.

Evaluate a System's Potential Capabilities

When determining the risks a specific system might pose to privacy, the system's *potential* capabilities must be assessed in addition to the risks involved in its intended usage. Few systems are deployed whose initial design and purposes are openly privacy invasive. Instead, systems may have unused capabilities, such as the ability to perform 1:N searches or be used with existing databases of biometric information, which could have an impact on privacy. Best Practices require that the impact of the deliberate misuse of a biometric system be considered when assessing whether a deployment is privacy invasive, neutral, or sympathetic.

This Best Practice poses challenges to both biometric system designers and privacy advocates, as it is difficult to determine how valid a potential capability for misuse must be to warrant attention. To illustrate, it is possible that a private-sector facial-scan database populated with customers may be compromised and searched in a 1:N fashion. While the risk of loss of personal data would be unlikely due to separation of biometric and nonbiometric data and the inherent limitations of facial-scan technology, it is possible that, over time, new facial-scan technologies will be developed that are better able to perform precise 1:N searches. Is this system, then, potentially privacy invasive? The answer depends on how essential the system is to a public- or private-sector organization, and who defines terms such as *potential* and *risk*. Privacy advocates are often unconcerned with how a biometric system is designed to operate and are instead concerned with how a system might be misused, whether innocently or maliciously.

Although systems with the potential to be used in a privacy-invasive fashion can still be deployed if accompanied by proper precautions, their operations must be monitored and protections must be in place to prevent misuse by internal or external parties. In extreme cases, the potential dangers of a compromised system may be so significant as to preclude its deployment, depending on the nature of the biometric implementation.

Limit Storage of Identifiable Biometric Data

Whenever possible, biometric data in an identifiable state, such as a facial image, fingerprint, or vocal recording, should be stored or used in a biometric system only for the initial purposes of generating a template. After template generation, the identifiable data should be destroyed, deleted, or otherwise rendered useless. This is to prevent the storage of fingerprints and facial images, as opposed to finger-scan and facial-scan templates. Templates are resistant to misuse because they cannot be identified as biometric information and cannot be used to re-create original biometric information.

Forensic systems and some public-sector programs store identifiable data in order to resolve borderline matches; in addition, employee background screens, which involve the acquisition of multiple fingerprint images, store identifiable data for future processing or auditing purposes. In this type of system, physical access and operational controls are necessary to ensure that identifiable data cannot be compromised.

Limit Collection and Storage of Extraneous Information

The nonbiometric information collected for use in a biometric verification or identification system should be limited to the minimum necessary to make identification or verification possible. Biometric databases generally comprise an index and a biometric template, with direct or indirect links to other databases as necessary. Storing names or account information is not only bad database design—this data will normally exist elsewhere and does not need to be collected and stored again—but also significantly increases the likelihood that biometric data may be associated with other personal information.

Make Provisions for System Termination

A method must be established by which a system used to commit or facilitate privacy-invasive biometric matching, searches, or linking can be depopulated and dismantled. The responsibility for making such a determination would rest with an independent auditing group and would be subject to appropriate appeals and oversight. This protection would apply primarily to public-sector systems, as they are most likely to be used in a privacy-invasive fashion and are more in need of independent oversight and monitoring. By contrast, private-sector deployments found to be privacy invasive will likely be modified or terminated as the result of pressure from investors, consumers, and the general public.

IBG BioPrivacy Best Practices: Data Protection

There are various ways to protect biometric systems, along with the data they store and transmit. These protections ensure that even if privacy-sympathetic biometric systems are attacked or compromised, the risk to privacy is mitigated.

Use Security Tools and Access Policies to Protect Biometric Information

Biometric information should be protected at all stages of its life cycle, including storage, transmission, and matching. The protections enacted may include encryption, private networks, secure facilities, administrative controls, and data segregation. The protections necessary within a given deployment are determined by a variety of factors, including the location of storage, the location of matching, the type of biometric used, the capabilities of the biometric system, whether processes take place in a trusted environment, and the risks associated with data compromise.

The protection of biometric information illustrates how security and privacy are related in biometric systems. Security is one of several tools—along with auditing, oversight, and disclosure—that increase user privacy by reducing the likelihood and impact of data compromise. It is important to note that security and privacy, though related, are not identical. As we've seen through the course of this chapter, privacy involves consent, nonlinkage, and individual control of personal data. Security relates to data confidentiality, integrity, and nonrepudiation—all tools essential to privacy protection, but by no means constituting the entirety of privacy protection.

Protect Postmatch Decisions

Data transmissions resulting from biometric comparisons should be protected to prevent replay attacks or compromise of personal information. When a successful biometric match takes place, this match is normally transmitted to the application or resource that requires authentication. This match decision must be protected, most likely through encryption, to avoid compromise of account information and to prevent man-in-the-middle attacks. Although these postcomparison decisions do not normally contain any biometric data, transmissions resulting from biometric matches are also sensitive. This protection is especially important in nontrusted environments such as the Internet.

Limit System Access

Compromise and unauthorized use of personal information is more likely to come from within an institution or organization than from outside. While numerous protections can be built to ensure that hackers cannot penetrate a biometric database from the Internet, it is a greater challenge to ensure that internal access to biometric systems is limited and controlled. In a worst-case scenario, an individual familiar with an institution's data reserves and system architecture could compromise a great deal of personal information without being detected.

Accordingly, access to biometric system functions and data must be limited to authorized operators and specific, controlled functions. A small set of trusted and accountable individuals should have the ability to update or view biometric databases, and then only along with controls on what data can be viewed and exported. Multiple-user authentication can be required when accessing or exposing especially sensitive data, meaning that more than one individual would need to collude to compromise sensitive data. Any employee access to databases that contain biometric information should be subject to controls and strong auditing. Limiting and monitoring system access is one of the most essential controls an institution can implement to ensure the privacy of individual data.

Implement Logical and Physical Separations between Biometric and Nonbiometric Data

Biometric data must be stored separately from personal information such as name, address, and medical or financial data. If biometric information is somehow compromised—if a template is intercepted or stolen from a hard drive, for example—protections must be in place to ensure that this cannot lead to compromise of other sensitive data.

In order for biometric data to be useful, it must be directly or indirectly associated with an identity, rights and privileges, account information, or some other piece of meaningful data. These associations, while a sine qua non of authentication systems, do result in increased risk that information can be linked to individuals. Creative ways of addressing the association problem—such as encrypting relational databases that contain pointers to account information, or developing pseudonym identifiers—can be implemented in situations where the risks of linking data are highest.

BioPrivacy Best Practices: User Control of Personal Data

Giving users control over how they can interact with a biometric system and how their data can be used reduces the risk that a biometric system will be perceived as privacy invasive.

Make System Usage Voluntary and Allow for Unenrollment

A basic privacy principle is that individuals have the right to control usage of their biometric information and can have it deleted, destroyed, or otherwise rendered unusable upon request. This extends to allowing individuals to decide whether to enroll in a biometric system and whether to continue in a system in which they have enrolled.

In customer-facing systems, the ability to opt out should not be problematic, as these systems will almost always be voluntary. Policies can be enabled that allow for users to employ standard authentication methods if desired. However, in certain public-sector and employment-related applications there is a compelling interest for the biometric system to be made mandatory and for biometric data to be retained for verification or identification purposes. In these cases, allowing users to opt out would render the system inoperable. When implementing a biometric system where usage is mandatory and ongoing, more expansive protections against misuse are necessary.

Enable Anonymous Enrollment and Verification

Depending on operational feasibility, biometric systems can be designed such that individuals can enroll and verify with varying degrees of anonymity. In Web environments, where individuals assume identities through email addresses or usernames, there may be no need for a biometric system to know with whom it is interacting, so long as the user can verify his or her original claimed identity. This identity may be associated with a number of purchases, the results of an anonymous medical test, or simply a series of communications. The key is that anonymous enrollment allows users to establish a claimed identity with a significantly reduced risk of profiling.

This Best Practice is consistent with the concept of limiting collection and storage of extraneous information. In effect, a name can be considered extraneous information in some implementations. Because biometric data is more difficult to track and associate than almost any other identifier—name, social security number, and so on—it can be effective at allowing anonymous enrollment and verification.

Provide Means of Correcting and Accessing Biometric-Related Information

System operators should provide a method for individuals to correct, update, and view stored information that is associated with biometrically enabled accounts. Just as consumers have access to their credit data and employees have a right to review files for accuracy, individuals interacting in any capacity with a biometric system have a right to review and contest information associated with their account. This is especially important because false matching may result in unauthorized access to a biometrically protected account, unbeknownst to the account holder.

In some cases, the ability to correct biometric-related information may extend so far as to allowing for reenrollment for users who wish to use a different biometric passphrase or sample. Such reenrollment would not be feasible in public-sector identification, but in 1:1 systems it would offer a greater degree of control to individuals.

IBG BioPrivacy Best Practices: Disclosure, Auditing, and Accountability

Even with the aforementioned protections in place, a system can be privacy sympathetic only if independent, objective parties review system operations and protections. This third-party review, as well as the disclosure that accompanies it, ensures that biometric systems operate in an open fashion, making them much less susceptible to misuse.

Make Provisions for Third-Party Auditing and Oversight

The single most critical protection against misuse and compromise of personal data is independent auditing and oversight. While any biometric deployer and vendor can claim to have developed privacy-sympathetic biometric systems, unless such claims are substantiated they are of little value. Trusted independent parties with authority to penalize breaches and enforce rules are required to ensure that institutions manage their biometric data in accordance with privacy principles. Depending on the nature of a given deployment, this independent auditing body ensures adherence to standards regarding data collection, storage, and use.

It is difficult to imagine a biometric system that could not somehow be used in privacy-invasive fashions, regardless of its intent. Instead of attempting to do the near impossible—to design a system that absolutely cannot be used for purposes beyond its original intent—it is more rational to build in as many protections as possible and to have the system monitored and audited for appropriate usage. Even if a perfectly privacy-sympathetic system were deployed, this third-party auditing and oversight would be necessary to ensure compliance.

Hold Operators Accountable for System Use and Misuse

The operators of certain biometric systems, especially large-scale systems or those deployed in the public sector, must be held accountable for system misuse, whether by internal or by external sources. In addition to motivating deployers to maintain privacy-sympathetic systems, accountability will increase the ability of auditors to enforce any judgments rendered on noncompliance.

Accountability is one of the few effective ways to combat scope creep. The temptation to expand the usage of biometric systems into unauthorized areas can be significant; the temptation to expand system usage will be reduced when decision makers in charge of system operations are held responsible for the performance of their systems and those who operate them. Along these lines, it should be clearly stated who is responsible for system operation, to whom questions or requests for information are addressed, and what recourse individuals have to resolve grievances.

Fully Disclose Audit Findings

Individuals should have access to findings gathered through third-party audits of biometric systems in order to facilitate public discussion on the system's privacy impact. These audits should present what protections are in place to prevent privacy-invasive usage (without risking security breaches), any issues encountered since the last audit, and what steps have been taken to address privacy-related security problems. As more light is shone on biometric deployments and their operations and policies become transparent, the likelihood of privacy-invasive usage is reduced.

Disclose the System Purpose and Objectives

The purposes for which a biometric system is being deployed must be fully disclosed to operators, enrollees, and any parties involved in the authentication. For example, is a system designed to locate and deter multiple enrollments, or is it being used for 1:1 verification? If individuals are told that the system will be used for identity verification, its usage in $1:N$ identity determination should not be permitted. By disclosing the purposes for which the system is being deployed, there will be less ambiguity should issues of scope creep arise.

The unauthorized distribution or use of biometric information is a type of function creep that could have severe consequences. Biometric information can easily be shared among corporations, partners, and other agents who may not adhere to the same data or privacy protections as the original entity to whom the user provided biometric data. Because of this, there should be no sanctions applied to any user who does not agree to broader usage of his or her biometric information. At the same time, as control over personal information is a basic privacy principle, individuals should have the option of extending the utility of their biometric data. For example, a user who accesses an online account through a biometric should be allowed the option of securing additional accounts with the same biometric information.

Disclose When Individuals May Be Enrolled in a Biometric System

Some biometric systems can enroll users without their knowledge or consent. Even if the enrollments are anonymous—that is, not associated with any specific identifying data such as name—individuals must be aware that biometric enrollments are being generated in a given area or through a recording device. Undisclosed enrollment in a facial-scan or voice-scan system reduces individuals' ability to consent to a system, as awareness is a precondition of consent. This disclosure may take the form of signage or a recording, depending on the biometric technology.

This poses challenges for biometric systems that enroll users from static images such as driver's license systems. If users are enrolled after the fact from static images, they are unable to consent to usage and are uninformed of system objectives. Disclosure must take place even if the enrollment templates are not being permanently stored, such as in a monitoring application. A basic privacy principle is that individuals must explicitly consent to the collection, use, and storage of personal information. Biometric data is personal information, and its surreptitious collection for the purpose of enrollment is inconsistent with reasonable privacy expectations.

Disclose When Individuals May Be Verified in a Biometric System

Similarly, ample and clear disclosure must be provided when individuals are in a location or environment where biometric matching (either 1:1 or 1:N) may take place without their explicit consent. This includes facial-scan systems used in public areas, keystroke-scanning technologies, signature-scan, and voice-scan—all technologies that can leverage existing technology to acquire biometric data. This notification can take many forms as long as individuals are informed of the system's operations. Should they choose not to interact with the biometric system, there should be reasonable accommodations made—not deactivating the system, of course, but perhaps allowing for a different type of authentication when possible.

Disclose Whether Enrollment Is Optional or Mandatory

Ample and clear disclosure must be provided that indicates whether enrollment in a biometric system is mandatory or optional. If optional, alternatives to the biometric should be made readily available. Individuals must be fully aware of their authentication options: There should be no implication that enrollment in a given system is compulsory if it is optional. Mandatory systems are, in general, subject to more rigid oversight because the risks of misuse are greater.

Disclose Enrollment, Verification, and Identification Processes

Individuals should be informed of the basic process flow of enrollment, verification, and identification, in order to provide them with a general understanding of how the system works. This includes detailing the type of biometric and nonbiometric information they will be asked to provide, the results of successful and unsuccessful positive verification, and the results of matches and nonmatches in identification systems. Furthermore, in 1:N systems where matches may be resolved by human intervention, the means of determining match or nonmatch should be disclosed. For example, if a forensic examiner is on staff to visually match two fingerprints, this information should be made available to enrollees. Though many individuals may not be interested in system operations or process flow, full disclosure is in the best interests of system operators.

This type of disclosure includes fallback procedures—those processes in place when biometric systems fail to verify authorized users or fail to enroll willing users, or when users choose not to enroll in a given biometric system. In order for individuals to make informed decisions about whether to enroll in a biometric system or what information to enroll, they should be aware of other authentication options, if any. These fallback procedures cannot be punitive or discriminatory in nature, although some biometric systems must be made mandatory to be effective. In a network security implementation, fallback procedures may include tokens, passwords, or smart cards; in physical access systems, users may have access to cards or PINs.

Disclose Policies and Protections in Place to Ensure Privacy of Biometric Information

Individuals should be informed of the protections used to secure biometric information, including encryption, private networks, secure facilities, administrative controls, and data segregation. Though it would be counterproductive to reveal so much information that the system is less robust or more subject to compromise, deployers should make key privacy protections available for review. Openness and transparency in these areas will increase confidence in system operations.

Biometrics at the Super Bowl: An IBG BioPrivacy Assessment

During Super Bowl week of January 21, 2001, a facial-scan system was deployed at Raymond James Stadium in Tampa Bay, Florida. The system, positioned at turnstiles in the complex, acquired facial images of event attendees and compared them against a database of "known felons, terrorists and con artists provided by multiple local, state and federal agencies." In this type of implementation, possible matches against watch lists are flagged so that law enforcement officials can make a manual determination as to whether there has actually been a match. The objective of the deployment was to increase public safety at this event.

When this implementation was announced to the public—after the Super Bowl—reaction was strongly negative. The event was referred to derisively as the "Snooper Bowl" because of the use of facial-scan, and it drew criticism from a range of privacy advocates. Even the mainstream press reported on this usage of biometrics, with the large majority of coverage associating this implementation with Big Brother surveillance fears.

When assessing the Super Bowl deployment according to the BioPrivacy Impact Framework, it is clear that a number of elements increased the potential privacy risk (see Table 16.1).

As this assessment through the BioPrivacy Impact Framework shows, several characteristics of this deployment are associated with increased privacy risk. To counter this, the following steps would be necessary to reduce the potentially privacy-invasive impact of this deployment:

➢ Full and open disclosure of the system's proposed usage prior to deployment
➢ Clear, explicit signage positioned to inform users prior to system interaction
➢ Protections against storage and/or misuse of collected data
➢ Full system oversight and auditing by independent parties
➢ Verification of nonretention of data
➢ Verification of system dismantling after the event
➢ Disclosure of criteria used to determine matches
➢ Penalties for noncompliance with the preceding minimum protections

Table 16.1 Biometrics at the Super Bowl: A BioPrivacy Impact Assessment

Overt ←——————→ Covert
Risk: 9/10
Although the acquisition devices (cameras) may have been in plain view, the fact that automated recognition technology was in use was not made clear. The biometric element, then, was covert. From a privacy perspective, this type of usage is more likely to become problematic because users are unaware of system operation.
Opt-in ←——————→ Mandatory
Risk: 8/10
The system was mandatory inasmuch as entry into the complex/facility required passage through a biometrically monitored turnstile. The ability to opt out is seen as a privacy benefit, but was not present in this environment.
Verification ←——————→ Identification
Risk: 9/10
Surveillance applications, by definition, are identification applications—the user is not claiming an identity, and the user's biometric data is compared against a database in order to locate a match.
Fixed Duration ←——————→ Indefinite Duration
Risk: 3/10
The system was in place for the week of the Super Bowl and then removed. The fixed duration is beneficial from a privacy perspective.
Private Sector ←——————→ Public Sector
Risk: 8/10
The system was used by local, state, and federal officials to conduct searches for known "felons, terrorists, and con artists." Although law enforcement applications are nearly always in the public sector, this does increase the risk that data could be shared across government bodies.

Table 16.1 Biometrics at the Super Bowl: A BioPrivacy Impact Assessment *(continued)*

Individual - Customer ←——|——→ Employee - Citizen

Risk: 2/10

The users whose facial-scan data was compared were under no compulsion to attend the event and were effectively acting in the capacity of customers. All other factors being equal, the use of biometrics in a customer environment, where coercion is minimal if at all existent, is less likely to pose a major privacy threat.

User Ownership ←——|——→ Institutional Ownership

Risk: 5/10

For the period during which the user's data was compared, the data was institution owned. On the other hand, the data was discarded unless the search resulted in a match, so the duration of ownership was limited. On the whole, the privacy impact was moderate.

Personal Storage ←————|————→ Template Database

Risk: 7/10

In this type of application, data is stored and processed in a centralized fashion. The fact that biometric templates were discarded after comparison is a mitigating factor.

Behavioral ←——|——→ Physiological

Risk: 5/10

Because physiological characteristics are less subject to change than behavioral, they are less contingent on user consent and cooperation. The use of facial-scan, which is a comparatively indistinct and less accurate physiological biometric, is a mitigating factor.

Templates ←——|——→ Identifiable Data

Risk: 3/10

The acquisition of biometric data was not the result of a tangible action of the user, but of the user's position within the facility. Deployments in which the user does not define the acquisition event are more likely to have a negative impact on privacy.

Conclusion

Biometrics are not inherently privacy invasive, although biometric systems can be deployed in privacy-invasive fashions. Adherence to Best Practices limits the harm that biometric systems can do to privacy, increases awareness of valid privacy concerns, and allows biometric systems to provide the benefits for which they are known—including increased security and convenience—without reducing privacy.

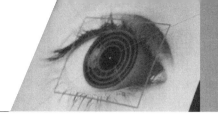

CHAPTER 17

Biometric Standards

The lack of industrywide standards has impeded many types of biometric implementations and has slowed the growth of the biometric industry. The relative youth of biometric technology, as well as the fragmented nature of the biometric industry, has resulted in sporadic and frequently redundant standards development. The only segment of the biometric industry with mature and widely adopted standards is live-scan fingerprint imaging, driven by the urgent need of law enforcement agencies. In the absence of biometric standards, some institutions have been wary of being locked into technologies that they perceived as being immature or developmental; standardization is taken as a sign of a technology's maturity.

The biometric industry, however, is actively addressing the standards problem, with some key efforts finalized and the process of industry adoption under way. Completed and ongoing standards efforts address a range of technical areas: application programming interfaces, file formats, encryption, image capture, device interoperability, and data exchange. This chapter discusses the most relevant biometric standards efforts and why they are important to deployers. It also identifies areas that biometric standards do not, and likely will never, address.

Why Standards?

At this stage in their development, the large majority of biometric systems, both hardware and software, are proprietary in many respects:

- The manner in which biometric devices and systems communicate with applications
- The method by which features are extracted from a biometric sample such as a fingerprint
- The method by which biometric data is compared
- The length and content of biometric templates, including header data and methods of encryption
- The method by which biometric data is stored and retrieved

The result of this is that, once a company decides to implement a certain biometric technology or to integrate biometric functionality into one of their products, in most cases they are wedded to that specific technology. Incorporating new technology requires that they rebuild their system from the ground up, in some cases replicating much of the deployment or development effort. Because of the emerging nature of the biometric industry, many potential biometric developers and deployers have delayed working with a specific technology for fear of their efforts being wasted if the biometric company ceases operations.

The widespread adoption of biometric standards will *not* make biometric technologies interoperable to the extent that a new device can replace an old device without reenrollment. The core algorithms by which vendors locate and extract data are unlikely to become standardized or interoperable, as these represent the basis of most vendors' intellectual property. With the widespread adoption of standards, once deployers and developers have integrated or implemented a biometric technology, they will be able to add a range of different biometrics much more easily. The risks of deploying or developing biometric solutions are significantly reduced. In addition, the development of standards in areas such as encryption and file formats ensures that the basic building blocks of biometric data management have been vetted in a collaborative fashion by industry professionals. Standards ensure that, in the future, biometric technology will be developed and deployed in accordance with generally accepted principles of information technology.

Application Programming Interfaces

The manner in which developers write applications that enable biometric solutions is clearly a fundamental issue for the biometric industry. Application programming interface (API) standards ensure that developers can address a wide range of biometric technologies and devices in a standardized fashion. The development of biometric APIs has been a long and contentious process, marked by competing efforts, mergers, alliances, and a major licensing agreement.

BioAPI

The BioAPI consortium has been one of the most prominent standards efforts since its inception in April 1998. With the support of major nonbiometric companies such as IBM, Hewlett-Packard (HP), and Compaq, BioAPI was developed as a replacement for alternative API initiatives that predated it, such as HA-API (which merged with BioAPI in March 1999). Formed to develop a "widely available and widely accepted API that will serve for various biometric technologies," the organization's stated goals are to work with industry biometric solution developers, software developers, and system integrators to leverage existing standards to facilitate easy adoption and implementation, to develop an OS-independent standard, and to make the API biometric-independent.

Broadly speaking, BioAPI is concerned with standardizing the way applications communicate with biometric devices and the way the data is manipulated and stored. It is not concerned with standardizing the way data is captured on the devices. This would allow companies that deploy BioAPI-compliant technologies to change devices without having to rewrite their software (although in this situation they would still need to reenroll users, as the underlying matching functions are not interoperable). BioAPI does this by giving application developers a common set of function calls for biometric devices. BioAPI is attempting to create "modular access to biometric functions, algorithms and devices"—a framework allowing programmers to develop once for a biometric device then easily make their work compatible with other devices. The capabilities addressed by the API are purposefully rudimentary: enrollment, verification, identification, capture, process, match, and store. The BioAPI framework is not intended to intrude upon the distinctive features that define each vendor's technology. Conversely, as stated in draft version 1.0, the goal is ". . . [t]o hide to the degree possible, the unique aspect of individual Biometric Technologies, and particular vendor implementations, products, and devices, while providing a high-level abstraction that can be used within a number of potential software applications."

The BioAPI consortium released its beta specification in September 2000 and its final specification in March 2001. While, as of this writing, very few solutions are officially BioAPI compliant—and the compliance program itself may not be finalized—BioAPI enjoys fairly broad acceptance within the biometric industry. The U.S. government has cited BioAPI compliance as a requirement for a handful of biometric-related projects, as government agencies played an integral role in the standard's development. Perhaps attributable to the number of companies and agencies involved in BioAPI development, BioAPI has had a prolonged development—it took years to produce a version 1.0, which in itself leveraged a preexisting standard. The time to maturity may have played a role in a major event that called into question the long-term relevance of BioAPI—namely, Microsoft's licensing of a competing standard: BAPI.

BAPI

From April 1999 to May 2000, BioAPI was the primary biometric API effort, with dozens of biometric vendors joining its development effort. A biometric API referred to as BAPI—which predated BioAPI, was merged with BioAPI, and formed the basis of some of BioAPI's underlying elements—was licensed by Microsoft in May 2000 for incorporation into future versions of its operating system. As opposed to being a consortium-based standard, BAPI was developed and is owned by I/O Software, a biometric middleware vendor.

Microsoft's licensing of BAPI came as a surprise to many in the biometric industry, as Microsoft had been an early supporter of BioAPI. However, the inclusion of biometrics as a core component of a Microsoft OS has helped legitimize the biometric industry, shifting the perception of biometrics from that of a futuristic technology to one that is becoming an everyday technology. Including biometric functionality in the operating system means that the OS will be capable of communicating with biometric devices in a standardized fashion. An analogy to printers is instructive: Before Microsoft standardized the way the OS communicates with printers, configuration and setup were arduous processes. Now, printer installation is greatly simplified, requiring only driver installation.

It is uncertain which version of the OS will include biometrics, but Microsoft has consistently reiterated its commitment to biometrics as a solution to authentication problems in network and e-commerce environments. Since the Microsoft announcement, I/O Software has also licensed BAPI elements to Intel for inclusion into its mobile PC security platform; this further solidifies BAPI's role in the biometric industry.

Deployers, Developers, and Biometric APIs

The long-term effects of the Microsoft announcement on the biometric industry are still unknown but are expected to be profound. It is possible that there will remain two competing efforts, with BAPI prevalent in the Windows/Intel arena and BioAPI gaining acceptance in U.S. government applications. Assuming that the Microsoft and Intel licensing and incorporation of BAPI continue as planned, it will be difficult for developers to overlook BAPI when planning for compatibility with biometric technologies. Biometric vendors within the BioAPI camp have also found themselves working increasingly with BAPI. On the other hand, for developers focused on customized government biometric implementations, BioAPI has already established itself as the API of preference.

File Format

The manner in which biometric templates are identified and exchanged is another variable that can impede or facilitate biometric development. A standard known as Common Biometric Exchange File Format (CBEFF) has been developed that defines standardized template formats, allowing systems to access and exchange different types of biometric data in a standardized format. CBEFF is a header format with optional and mandatory fields that define common elements for exchange between biometric devices and systems, such as data security options, data integrity options, biometric type, creation date, and signature. The standard does not provide device or matching interoperability, but does provide a common method of handling biometric data. As stated on the Biometric Consortium's Web site, "CBEFF data can be placed in a single file used to exchange biometric information between different system components or between systems."

CBEFF has been aligned with BioAPI and X9.84 (see next section), and as such forms the basis of more developer- and deployer-facing standards efforts. CBEFF is relevant primarily to biometric vendors, who must ensure that the header files that precede biometric data in a biometric template are CBEFF compliant. The standard has been finalized and is being migrated to environments such as smart cards.

Information Security for Financial Services

The financial services industry frequently requires that standardized methods be implemented in their systems and processes. X9 is the organization responsible for developing and publishing voluntary consensus technical standards for the financial services industry. Within X9, a biometric standards effort known as *X9.84-2000 Biometric Information Management and Security* has been completed and addresses security and cryptography of biometric data and biometric systems. X9.84's importance comes from the financial services industry's desire to have standards implemented before investing significant resources in a new technology.

Accredited by American National Standards Institute, X9 represents banks, credit unions, government regulators, equipment manufacturers, investment managers, and the like. X9 addresses a range of financial services applications, as described in their Web site: "Check processing, electronic check exchange, PIN management and security, financial industry use of data encryption, and wholesale funds transfer, among others. Standards under development include electronic payments on the Internet, financial image interchange, home banking security requirements, institutional trade messages, and electronic benefits transfer." Clearly these areas are of significant interest to any company providing logical access via biometrics.

X9.84 is concerned with the security and management of biometric data across its life cycle, including secure transmission and storage, and security of hardware. X9.84 also attempts to address verification and identification of bank employees and customers. Their focus is on the integrity of the overall system—they view biometric data as a type of public key whose disclosure "should not compromise the system or the individual."

Functionally, the X9.84 standard plans to accompany the architecture BioAPI is developing. BioAPI's standardized header accompanies the raw biometric data, and X9.84 then defines the types of protections that systems deployed in banking environments should implement. These protections can include technologies such as encryption, designed to provide integrity, nonrepudiation, and confidentiality. X9.84 has reviewed the entire process flow of the biometric authentication process for security risks, weak points, and so on. Their logic is that if the system is strong, then banks and similar institutions can implement with confidence in the source of the data, the verification results, the integrity of the data, the confidentiality of the data, the transmission of the data, and the storage of the data, as well as any other variable that may arise.

The X9.84 standard was approved in March 2001 and is also planned for submission to the ISO as an ANSI standard thereafter. Such standards are voluntary, of course, but must be adhered to if the institutions considering implementation require it, as is very likely. While X9.84 will provide banking institutions with increased assurance that biometrics can be implemented effectively, it may complicate the task of biometric vendors, especially those who perform the extraction, comparison, and transmission functions. These vendors may need to incorporate a level of security and encryption far beyond what they implement currently.

Additional Efforts

In addition to the aforementioned key standards, other efforts at standardization that may be important to the future of biometrics are under way.

Fingerprint Template Interoperability

An ANSI-certified committee relevant to biometrics is the National Committee for Information Technology Standards (NCITS), pronounced "insights." Formerly known as X3, NCITS's mission is to "produce market-driven, voluntary consensus standards in the areas of multimedia, intercommunication among

> ## Background: NIST, ITL, ISO, and ANSI
>
> Some background on agencies dedicated to standards development and approval may be helpful. The National Institute of Standards and Technology (NIST), an agency with the U.S. Department of Commerce's Technology Administration, was established by Congress to "assist industry in the development of technology needed to improve product quality, to modernize manufacturing processes, to ensure product reliability and to facilitate rapid commercialization of products based on new scientific discoveries." A division of NIST, the Information Technology Laboratory (ITL), performs testing, testing methods, and proof-of-concept implementations.
>
> The International Organization for Standardization (ISO) is the primary nontreaty worldwide standardization organization. The standards bodies of 130 countries are represented in this nongovernmental federation. Developing standards in such varied fields as textiles, small craft, and cleanrooms, ISO is committed to "facilitating the international exchange of goods and services, and to developing cooperation in the spheres of intellectual, scientific, technological and economic activity." The U.S. representative to ISO is the American National Standards Institute (ANSI), an organization that confers accreditation to qualified groups in various fields for the purpose of developing American National Standards (ANSs).

computing devices and information systems, storage media, database, security, and programming languages." Their documents are frequently referred to as "ANSI NCITS." NCITS consists of approximately 35 committees, one of which is B10, dedicated to developing standards for identification cards and related devices. A primary working group of B10 is B10.8, dedicated to driver's licenses and similar identification cards. A task force within B10.8 developed a very interesting biometric standard defining a common method of extracting and processing features from a fingerprint image. This interoperability would enable states to share biometric data for the purposes of driver's license verification.

Standardization among driver's license applications is a very challenging but potentially lucrative area of biometrics. Interoperability would enable cross-vendor 1:1 matching for those who adopt the standard. This feat is currently impossible: Vendors closely guard the proprietary methods by which they extract such minutiae. This lack of interoperability has hampered the development of truly large-scale rollouts, as few states or organizations are eager to fund projects incorporating proprietary technology.

Standardization at the matching level is complicated by the variables involved in feature extraction and comparison. Feature extraction for fingerprints generally refers to marking the distinctive points (minutiae) on a fingerprint where ridges end or split. Both ridge endings and bifurcations have several subcategories, and some vendors break these out separately. Some vendors rate minutiae quality. The way minutiae are situated differs from vendor to vendor—some base measurement on distance from the core; others, relative to position on the platen. Vendors occasionally divide the fingerprint into sectors and count minutiae therein; others count ridges between minutiae. Clearly, establishing a bedrock extraction methodology, however simplified, will involve much work and compromise on the part of biometric vendors.

CDSA/HRS

The Common Data Security Architecture Specification/Human Recognition Services (CDSA/HRS) architecture is partially involved in biometric standards development. Begun by Intel in December 1997, CDSA/HRS is developing a secure, multiplatform software framework for applications including "electronic commerce, communications, and digital content." They are working directly with the BioAPI consortium to maximize consensus on a widely accepted framework. CDSA is developing a common API to which programmers can add authentication functionality (incorporating biometrics, for example). Members include Hewlett-Packard, IBM, Motorola, Netscape, Shell Companies, and JP Morgan. The HRS component is an extension to the CDSA architecture directly related to authentication. Considering the heft of CDSA contributors, the adoption of this architecture can only help the biometric industry as a whole.

Conclusion

The emergence of biometric standards will facilitate growth in the biometric industry and ensure that biometric solutions brought to market meet minimum standards for interoperability from a developer's perspective. Interestingly, corporations and interests outside the biometric industry have driven standardization as much as, if not more than, biometric vendors. These outsiders have often had a broader perspective on how standards are essential to the widespread distribution and acceptance of the technology than the biometric industry. In some cases, biometric companies have opposed standards out of fear that they might undermine the uniqueness of their product offerings. However, increased standardization should contribute to an increase in the size of the biometric playing market. Standardization does not mean that companies will have to compromise what makes their products unique, nor does it mean that there will not be room for companies that specialize in integration, training, configuration, support, and everything that helps convert quality products into successful projects.

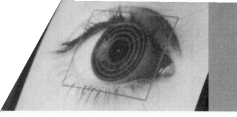

Index

1:1 (one-to-one), 12
1:1 systems enrollment, 17
1:few (one-to-few) identification, 14
1:N (one-to-N or one-to-many), 12
1:N identification, 26
  ATMs, 206
  BioPrivacy Impact Framework, 249–250
  enrollment, 17
  retail/ATM/point of sale (POS), 206

A

access policies, 265–266
accountability, 5, 267–272
  retina-scans, 106–107
  system use and abuse, 269–270
accuracy
  basic algorithm-level testing, 23
  changes, 28
  derived metrics, 38–40
  environmental changes, 29
  finger-scan, 58
  FMR (false match rate), 24–27
  FNMR (false nonmatch rate), 27–28
  FTE (failure-to-enroll) rate, 33–38
  user presentation changes, 29
active capacitance, 54
AFIP (Arizona Fingerprint Imaging Program), 120
AFIRM (Automated Fingerprint Image Reporting and Match), 120
AFIS (automated fingerprint identification systems)
  1:N matching, 120, 121
  central database storage, 121
  citizen identification, 159, 161
  components, 114–115
  criminal identification, 154
  data acquisition, 116
  deployments, 119–120
  distinctive features, 118
  enrollment, 116
  expensive live-scan device scanners, 121
  fingerprint images, 120
  finger rolled fully from nail to nail, 121
  vs. finger-scans, 120–121
  government sector, 214
  jurisdiction, 119
  large-scale identification, 118, 120
  law enforcement, 119, 211–212
  matching, 118–119
  penetration rate, 118
  public sector, 121
  travel and immigration, 229
  use of multiple fingerprints, 117
air travel, 229–230
Allied Iris Bank, 93
American Academy of Ophthalmology, 79
anonymous enrollment, 268
anonymous verification, 268
ANSI (American National Standards Institute), 281, 283
ANSs (American National Standards), 283
APIs (Application Programming Interfaces)
  BAPI, 280
  BioAPI, 279–280

285

Index

application-specific privacy risks, 246–255
Argentina, 119, 215
ATMs
  1:*N* identification, 206
  financial sector, 222–223
ATV (ability-to-verify) rate, 39–40
audio capture devices, 89, 90
auditing, 267–272
authentication, 12
  automated, 9
  biometric middleware, 57
  vs. biometrics, 3–5
  biometric software, 172–173
  e-commerce/telephony, 191, 194
  exclusivity, 149
  hand-held tokens, 3–4
  higher levels of rights and privileges, 5
  passwords, 3–4
  PC/network access, 179
  physical access/time and attendance, 186
  PINs, 3–4
  problem solution, 149
  receptiveness, 149
  retail/ATM/point of sale (POS), 207
  susceptible data, 173
  urgency of problem, 148
authorize transaction message, 124
automated DNA solutions, 152–153
automated fingerprint searches, 152
automated use, 9–10

B

B2B (business-to-business) e-commerce, 15
B2C (business-to-consumer) e-commerce, 15
background checks, 158
Bacob, 93
Bank United, 82, 222
BAPI, 174, 220, 280
basic algorithm-level testing, 23
behavioral biometrics, 10
behavioral characteristics, 10
Behavioral/Physiological category, 256–258
behavioral *vs.* physiological technology, 254
benefits distribution, 218–219
Ben-Gurion Airport, 102, 230
bifurcations, 51
BioAPI, 174, 220, 279–280
Biometrica, 232
biometric applications
  citizen identification, 145
  criminal identification, 144
  defining, 144–146
  e-commerce/telephony, 145
  emerging, 146
  horizontal approach, 143
  identify or verify identity, 144
  logical access, 144
  mature, 146
  PC/network access, 145
  physical access, 144
  physical access/time and attendance, 145
  privacy infringement, 147
  retail/ATM/point of sale, 144
  surveillance, 145–146
  usage, 147
biometric comparisons, 22
biometric data
  acquiring and comparing, 15
changes in, 28
feature extraction, 18
presentation, 17
reconstructing from templates, 19
biometric hardware options, 57
biometric identification, 5
biometric industry size, 196
biometric information
  correcting, 268
  function creep, 242
  limiting retention of, 261
  as personal information, 243
  unauthorized collection of, 241–242
  unauthorized disclosure, 242
  unauthorized use of, 240–241
  unnecessary collection of, 242
biometric matching
  decision, 21
  enrollment, 16–18
  match, 20
  nonmatch, 20
  process flow, 15–16
  score, 20–21
  template creation, 16–18
  threshold, 20, 21
biometrics
  accountability, 5
  benefits of, 3–5
  convenience, 5
  definition of, 9
  deployments and privacy, 238–239
  identification systems, 6
  layered solutions, 132
  metrics, 24
  noncommercialized technologies, 113
  point-of-sale transactions, 5

Index

privacy-sympathetic qualities, 244–246
security, 4–5
vs. traditional authentication methods, 3–5
as unique identifier, 260–261, 262–263
biometric sample, 17
Biometrics Management Office, 220
biometric software, 57
Biometric Solution Matrix, 147–148
  citizen identification, 160
  criminal identification, 155
  e-commerce/telephony, 195
  effectiveness, 149
  exclusivity, 149
  PC/network access, 175
  physical access/time and attendance, 184-185
  receptiveness, 149
  retail/ATM/point of sale (POS), 204–205
  scope of authentication problem, 149
  solution to authentication problem, 149
  surveillance, 166
  urgency of authentication problem, 148
biometric standards, 277
  APIs (Application Programming Interfaces), 278–280
  CDSA/HRS (Common Data Security Architecture Specification/Human Recognition Services), 284
  file formats, 281
  financial services information security, 281–282
  fingerprint template interoperability, 282–284
  reasons for, 277–278
biometric systems
  broken, 4
  convenience, 40
  cost, 39
  degrees of certainty, 40
  difficult for users to interact with, 29
  disclosing purpose and objectives, 270
  enrollment disclosures, 271
  facilitate proper presentation, 29
  imposter break-in, 25
  logical access systems, 14–15
  optional or mandatory enrollment, 271–272
  physical access systems, 14–15
  scoring, 20–21
  security, 40
  speed of, 9–10
  verification disclosures, 271
biometric templates, 18–20
BioPrivacy Best Practices, 246, 259
  accountability, 267–272
  auditing, 267–272
  biometrics as unique identifier, 260–261
  data protection, 265–267
  disclosure, 267–272
  extraneous data collection and storage limitations, 264–265
  identifiable data storage limitation, 264
  limiting retention of biometric information, 261
  limiting system scope, 260
  system potential capabilities evaluation, 263–264
  system termination provisions, 265
  user control of personal data, 267–268
BioPrivacy Impact Framework, 246
  behavioral *vs.* physiological technology, 254
  citizen, employee, traveler, student, customer, individual, 251–253
  fixed duration *vs.* indefinite duration, 250
  opt-in *vs.* mandatory enrollment, 248–249
  overt *vs.* covert collection, 246–248
  personal storage *vs.* template database storage, 253–254
  public sector *vs.* private sector, 250–251
  template storage *vs.* identifiable data storage, 255
  user ownership *vs.* institutional ownership, 253
  verification *vs.* identification, 249–250
BioPrivacy Technology Risk Ratings, 246, 256–258
booking stations, 72
border crossing, 230–231
Brazil, voter ID and elections, 216
British Airways, 83, 230
British Columbia Institute of Technology, 232
Buytel, 93

C

Canada
  border crossing, 230
Canadian Airports Council Expedited Passenger Processing System Project, 83, 230
CANPASS, 102
CBEFF (Common Biometric Exchange File Format), 174, 220, 281
CDSA/HRS (Common Data Security Architecture Specification/Human Recognition Services), 284
cepstral coefficients, 90
Charles Schwab & Co., 93, 127–128
Charlotte/Douglas International Airport, 83, 231
Charter Schools, 233
China and national IDs, 215–216
citizen authentication, 147
citizen-facing applications, 152
  centralized storage of biometric data, 151
  citizen identification, 157–164
  large-scale deployments, 151
  surveillance, 164–169
citizen identification, 145
  AFIS (automated fingerprint identification systems), 161
  AFIS systems, 159
  background checks, 158
  Biometric Solution Matrix, 160
  cost to deploy biometrics, 161–162
  deployment issues, 162–164
  driver's license or identification card issuance, 158
  effectiveness, 160
  enrollment logistics, 162
  error rates, 163
  exclusivity, 160
  facial-scans, 161
  finger-scans, 161
  future trends, 158–160
  government benefits programs, 159
  government entitlements or benefits, 157
  immigration-related activities, 158
  legacy systems, 163–164
  logistics, 160
  multifunction cards, 159
  privacy-sympathetic data handling, 164
  receptiveness, 160
  related biometric technologies and vertical markets, 161
  response times, 163
  scalability, 162–163
  scope, 160
  typical applications, 157–158
  urgency, 160
  voter registration, 159
  voting and voter registration, 157
citizens, 251–253
CMOS (complementary metal oxide semiconductor), 179
Compaq, 279
Comparative Biometric Testing, 25, 59, 127
comparison, 22
components
  facial-scan, 64
  finger-scan systems, 46–48
consumer authentication, 147
ContinuedEd.com, 232
contributory database surveillance, 232
convenience, 5, 40
correcting biometric information, 268
corrections applications and law enforcement, 213
cost, 39
Costa Rica voter ID and elections, 216
Credit Union Central, 173
criminal identification, 144
  1:N identification, 152
  AFIS technology, 154
  automated DNA solutions, 152–153
  Biometric Solution Matrix, 155
  costs of deploying biometrics, 156
  current applications, 152
  effectiveness, 155
  exclusivity, 155
  facial-scans, 154
  future trends, 152–153
  price, 152
  receptiveness, 155
  related biometric technologies and vertical markets, 154
  scope, 155
  urgency, 155
crossover rate, 38-39
current applications
  criminal identification, 152
  PC/network access, 172–173
  physical access/time and attendance, 181–182
  retail/ATM/point of sale (POS), 201–202
  surveillance, 164–165

Index

customer-facing
  applications, 189
  e-commerce/telephony, 189–201
  retail/ATM/point of sale (POS), 201–208
customers, 251–253

D

data acquisition
  AFIS, 116
  hand-scans, 101
  keystroke-scans, 134–135
  signature-scans, 125
  voice-scans, 89–90
data processing
  hand-scans, 101
  keystroke-scans, 134–135
  signature-scans, 126
  voice-scans, 90
data protection
  access policies, 265–266
  limiting system access, 266
  logical and physical separation between data, 267
  protecting postmatch decisions, 266
  security tools, 265–266
derived metrics, 38–40
desktop cameras, 78, 80
deterrence and hand-scan systems, 105
disclosure, 267–272
distinctive characteristics
  facial-scan, 67–68
  finger-scan, 51
DNA, 152–153
DNA identification, 113
DNA matching, 153
Dominican Republic voter ID and elections, 216
DPI (dots per inch), 48
Dresdner Bank, 82–83, 222
driver's license or identification card issuance, 158
Dutch Immigration and Naturalization Department, 231

E

e-commerce/telephony, 145
  acquisition devices, 192
  authentication, 191, 194
  Biometric Solution Matrix, 195
  costs, 193–194
  current applications, 190–191
  deployment issues, 198–201
  effectiveness, 195
  enrollment, 198
  exclusivity, 195
  facial-scans, 193
  fallback procedures, 199
  finger-scans, 192, 193, 199
  fraudulent enrollment, 198
  future trends, 191–192
  increased trust in remote transactions, 191
  integration into existing systems, 199–200
  internal or external infrastructure, 200–201
  iris-scans, 193
  lack of trust, 190
  liability, 198–199
  locking accounts, 199
  new devices and technologies compatibility, 199
  password authentication, 199
  potential restrictions, 192
  receptiveness, 195
  related biometric technologies and vertical markets, 193
  remote parties, 190
  scope, 195
  security, 200
  transactional revenue models, 191
  transaction usage, 190
  trusted parties, 192
  urgency, 195
  utilization of, 190–191
education, 232–233
EER (equal error rate), 38-39
effectiveness
  citizen identification, 160
  criminal identification, 155
  e-commerce/telephony, 195
  PC/network access, 175
  physical access/time and attendance, 184
  retail/ATM/point of sale (POS), 204
  surveillance, 166
electronic tablets, 131
emerging biometric applications, 146
employee authentication, 147, 219
employee-facing applications
  PC/network access, 171–180
  physical access/time and attendance, 180–187
employees, 251–253
enrollment, 16–18, 33
  AFIS (automated fingerprint identification systems), 116
  anonymous, 268
  difficult for users to provide data, 34
  disclosures, 271
  e-commerce/telephony, 198
  enrollment score, 21

enrollment *(continued)*
  finger-scan, 52
  fraudulent, 198
  identifier, 18
  keystroke-scans, 134–135
  logistics and citizen identification, 162
  low-quality, 198
  multiple enrollment templates, 18
  PC/network access, 179
  process disclosure, 272
  quality score, 21
  sharing data, 198
  signature-scans, 130
  varied processes, 34
enrollment score, 21
enrollment templates, 20, 21
  facial-scans, 68
  feature analysis, 71
enterprise applications, 173
entity authentication, 225
environment, changes in, 29
error rates for citizen identification, 163
E*Trade, 224
exclusivity
  citizen identification, 160
  criminal identification, 155
  e-commerce/telephony, 195
  PC/network access, 175
  physical access/time and attendance, 184
  retail/ATM/point of sale (POS), 204
  surveillance, 166
expanded service kiosks, 223
extraneous data collection and storage limitations, 264–265
eye-based biometrics, 28, 107
EyeTicket, 83

F

face location engine, 64
face recognition engine, 64
facial-scans, 10, 17, 29
  1:*N* identification, 26
  acquisition environment effect on matching accuracy, 74
  AFP (automatic face processing), 71
  away from ideal positioning, 65
  booking stations, 72
  changes in physiological characteristics, 74–75
  citizen identification, 161
  competing technologies, 69–72
  components, 64
  core technology, 64
  criminal identification, 154
  deployments, 72
  distance from camera, 65
  distinctive characteristics, 67–68
  driver's license system, 72
  e-commerce/telephony, 193
  effectiveness, 168
  Eigenface, 69–70
  enrolling static images, 73–74
  enrollment templates, 68
  face location engine, 64
  face recognition engine, 64
  facial changes, 68
  feature analysis, 70–71
  financial sector, 221
  FNMR (false nonmatch rates), 30
  FTE (failure-to-enroll), 35
  government sector, 214
  high-quality enrollment, 65
  high-resolution cameras, 65
  ID card applications, 72
  identification, 13
  image acquisition, 65–66
  image processing, 66
  law enforcement, 211, 212–213
  leveraging existing equipment and imaging processes, 73
  licensing, 167–168
  lighting, 66
  neural network, 71
  operating without physical contact or user complicity, 73
  PC/network access, 177
  privacy abuse, 75
  races and ethnicities, 66
  retail/ATM/point of sale (POS), 206
  strengths, 63, 72–74
  surveillance, 72, 167
  susceptible to changes, 28
  technology risk ratings, 257
  template creation, 68
  template matching, 64, 68–69
  weaknesses, 63, 74–75
facial-scan systems
  1:1 verification, 66
  1:*N* public-sector identification, 66
  identification comparisons, 69
  postmatch integration functionality, 64
  surveillance, 66
  temporal changes, 28
false matches
  acceptability, 25–26
  single comparison templates, 26
false minutiae, 52
false nonmatches and authentication environment changes, 29

Index

false nonmatch rate, 24
false verification attempts, 36
FAR (false acceptance rate), 25
FBI, 119
feature analysis, 70–71
feature extraction, 18
feedback, 29
file formats, 281
financial sector, 210, 220
  account access, 222
  ATMs, 222–223
  expanded service kiosks, 223
  facial-scans, 221
  finger-scans, 221
  hand-scans, 221
  iris-scans, 221
  online banking, 223–224
  PC/network access, 224
  physical access, 224
  retina-scans, 221
  signature-scans, 221
  technologies, 221
  telephone transactions, 224
  typical deployments, 222–224
  voice-scans, 221
financial services information security, 281–282
Financial Services Modernization Act, 224
finger imaging, 157
fingerprints, 117-118
  bifurcations, 51
  deltas, 51
  distinctive data, 118
  minutiae, 51, 52
  quality check on data, 118
  ridges and valleys, 51
  temporal changes, 28
fingerprint template interoperability, 282-284
fingerprint templates, 114

finger-scan devices, 117
finger-scans, 10, 17
  1:1 authentication, 120
  vs. AFIS (automated fingerprint identification systems), 120-121
  Asian populations, 60
  citizen identification, 161
  competing technologies, 54–56
  deployment environments, 58–59
  deployments, 56–58
  distinctive characteristics location, 51
  distortions, 52
  ease of use, 111
  easy-to-use devices, 59
  e-commerce/telephony, 192, 193, 199
  elderly populations, 59
  enrolling multiple fingers, 59
  enrollment, 52
  false match, 83
  false minutiae, 52
  financial sector, 221
  FNMR (false nonmatch rates), 30
  forensic applications association, 60
  FTE (failure-to-enroll), 35
  government sector, 214
  healthcare, 226
  high levels of accuracy, 58
  identification, 13
  image acquisition, 48–50
  image processing, 50–51
  inability to enroll some users, 59–60
  law enforcement, 213
  location of fingerprint, 50
  manual laborers, 60
  minutiae, 55–56
  modules, 47–48
  optical technology, 54

pattern matching, 55–56
PC/network access, 177, 178, 192
performance deterioration over time, 60
physical access/time and attendance, 183
privacy, 60
retail/ATM/point of sale (POS), 202, 206
silicon technology, 54–55
specialized devices, 60–61
storing data, 120
strengths, 45, 58–59
technology risk ratings, 257
template creation, 52
template matching, 52–53
templates, 120
thermal imaging, 55
travel and immigration, 229
ultrasound technology, 55
weaknesses, 45, 59–61
finger-scan systems
  components, 46–48
  finger-scan template, 120
  flat placement of fingerprint, 120
  home applications, 121
  inexpensive peripherals, 121
  integrated with external systems, 48
  matching, 48
  measurements, 47
  passwords, 48
  platen, 47
  private-sector, 121
  public-sector, 121
  scanner, 47
  template generation, 48
  user feedback, 29
fixed duration *vs.* indefinite duration, 250

FMR (false match rate)
 identification system, 24
 importance of, 25
 large-scale identification systems, 27
 minimizing, 25
 reduction, 24
 single, 26–27
 system, 26–27
 verification system, 24
FNMR (false nonmatch rate), 27
 factors affecting, 30–32
 importance of, 28
 large-scale identification systems, 33
 real-world, 31
 relation to FTE (failure-to-enroll), 36–37
 single, 31-32
 system, 31-32
forensic fingerprinting, 60
formant frequencies, 90
Foxwoods Casino, 72
fraud detection, 6
fraud deterrence, 6
FRR (false rejection rate), 27
FTE (failure-to-enroll), 33
 across different populations, 35-37
 dependence on system design and training, and ergonomics, 34
 importance of, 34–35
 large-scale identification systems, 38
 relation to FNMR (false nonmatch rate), 35–37
 single, 37–38
 system, 37–38
function creep, 242
future trends
 citizen identification, 158–160

criminal identification, 152–153
e-commerce/telephony, 191–192
PC/network access, 173–175
physical access/time and attendance, 182
retail/ATM/point of sale (POS), 203–204
surveillance, 165

G

gain, 90
gait recognition, 113
gaming, 232
Give/Grab category, 256–258
Glendale, California employee authentication, 219
Glendale Police Department, 56
good passwords, 4
government entitlements or benefits, 157
government sector, 210
 AFIS, 214
 benefits distribution, 218–219
 driver's licenses, 217
 employee authentication, 219
 facial-scans, 214
 finger-scans, 214
 military programs, 220
 national IDs, 215–216
 technologies used in, 214
 typical deployments, 215–220
 vertical markets, 214–220
 voter ID and elections, 216–217
Griffin Investigations, 72, 232

Groupo Financiero Banorte, 56, 223
GSA (General Services Administration) Common Access Card, 219
Gujarat driver's licenses, 217

H

hand-held tokens, 3–4
hand-scans, 10, 17, 99
 components, 100
 convenience, 104–105
 costs, 106
 data acquisition, 101
 data processing, 101
 deployments, 102
 deterrence, 104–105
 distinctive features, 101
 financial sector, 221
 FNMR (false nonmatch rates), 30
 FTE (failure-to-enroll), 35
 healthcare, 226
 operation challenging environments, 103
 perception as nonintrusive, 104
 physical access/time and attendance, 183
 physiological characteristic as basis, 104
 reliability, 104
 strengths, 99, 103
 susceptible to changes, 28
 technology risk ratings, 258
 template generation, 102
 template matching, 102
 travel and immigration, 229
 unchanged technology, 104
 verification, 101
 weaknesses, 100

Index

hand-scan systems
  deterrence, 105
  limited accuracy, 105
  scope of potential
    applications
    limitations, 105
  security, 105
  storing templates, 100
  weaknesses, 105–106
Harriet Beecher Stowe
  Elementary School, 233
Hawaii Airport, 231
healthcare, 210, 225
  finger-scans, 226
  hand-scans, 226
  iris-scans, 226
  patient identification,
    227–228
  PC/network access,
    226–227
  personal information
    access, 227
  technologies, 226
  typical deployments,
    226–228
Heathrow Airport, 83, 230
Hewlett-Packard, 279
Hidden Markov models,
  91–92
highly controlled environments, 23
HIPAA (Health Insurance
  Portability and
  Accountability Act),
  225, 239
Houston Municipal Credit
  Union, 222

I

IAFIS (integrated
  automated fingerprint
  identification
  systems), 115, 212
IBM, 279
ID card applications, 72
identifiable data storage
  limitation, 264
identification, 12–13
  appropriateness of, 13–14
  BioPrivacy Impact
    Framework, 249–250
  facial-scan, 13
  finger-scan, 13
  iris-scan, 13, 82
  process disclosure, 272
  retina-scan, 13
identification systems, 10,
  12
  biometrics, 6
  computational power, 13
  error rates, 6
  errors, 13
  false nonmatch, 27
  FMR (false match rate),
    24
  fraud detection, 6
  fraud deterrence, 6
  large-scale, 13
  large-scale public benefits
    programs, 13
  negative, 12–13
  positive, 12
identifier, 18
identity, 10, 11
  manually verify or
    determine, 9
  verification and determi-
    nation, 10-11
image acquisition
  environmental factors, 50
  facial-scan, 65–66
  finger-scan, 48–50
  iris-scans, 79–80
  retina-scans, 108
image processing
  facial-scan, 66
  finger-scan, 50–51
  iris-scans, 80
image quality, 48
Imagis, 232
immigration-related
  activities, 158
imposter break-in, 25, 26,
  36
inconclusive, 21
individuals, 11, 251–253
inexpensive desktop
  cameras, 79
informational privacy,
  239–243
infrared imager, 79
InnoVentry, 223
INSPASS (Immigration
  and Naturalization
  Service Passenger
  Accelerated Service
  System), 102, 230
Integrated AFIS, 156
Intel, 284
intensity, 90
International Biometric
  Group, 36
international legislatures
  employee authentica-
    tion, 219
I/O Software, 280
Iridian, 83
iris-scans, 10, 17, 107
  acquisition hardware, 78
  components, 78–79
  deployments, 82–83
  desktop cameras, 80
  desktop devices, 85
  difficulty of usage, 85
  distinctive features, 80
  ease of use, 111
  e-commerce/telephony,
    193
  false nonmatching and
    failure to enroll, 85
  false rejection, 85
  financial sector, 221
  FNMR (false nonmatch
    rates), 30
  FTE (failure-to-enroll), 35
  healthcare, 226
  identification, 13, 82
  image acquisition, 79–80
  image processing, 80
  inexpensive desktop
    cameras, 79

iris-scans *(continued)*
  infrared illumination, 78
  iris-pupil border, 80
  kiosk-based systems, 79
  law enforcement, 211, 213
  physical access devices, 79
  physical access/time and attendance, 183
  proprietary acquisition device need, 86
  reenrollment, 84
  resistance to false matching, 83–84
  retail/ATM/point of sale (POS), 206
  software components, 78
  stability of characteristic over lifetime, 84
  strengths, 77, 83–84
  suitability for logical and physical access, 84
  susceptible to changes, 28
  technology risk ratings, 257
  template generation, 81
  template matching, 82
  trabecular meshwork, 80
  travel and immigration, 229
  user discomfort, 85–86, 111
  verification, 82
  weaknesses, 77, 84–86
  Web-enabled, 78
ISO (International Standards Organization), 283
Israel and border crossing, 230
Italy voter ID and elections, 216
ITL (Information Technology Laboratory), 283

K

keystroke-scans, 17, 132–133
  changeable usernames and passwords, 137
  common authentication process, 136–137
  components, 134
  data acquisition, 134–135
  data processing, 134–135
  distinctive features, 135
  double-digit false matches, 138
  enroll and verify users with little effort, 137
  enrollment, 134–135
  enrollment template, 137
  existing hardware leverages, 136
  flaws of password-based systems and, 138–139
  FNMR (false nonmatch rates), 31
  FTE (failure-to-enroll), 35
  not increasing user convenience, 138
  one-to-one verification, 137
  passwords, 139
  privacy, 137
  security, 136–137
  single-sitting enrollment, 135
  strengths, 133, 136–137
  technology risk ratings, 258
  template generation, 135–136
  template matching, 135–136
  templates, 137
  unproven technology, 138
  usernames, 139
  verification, 134–135, 137
  weaknesses, 134, 137–139
kiosk-based systems, 79

L

land telephones, 89, 90
large-scale AFIS systems, 119
large-scale systems, 13
  FMR (false match rate), 27
  FNMR (false nonmatch rates), 33
  FTE (failure-to-enroll), 38
  verification systems, 33
law enforcement, 210
  AFIS, 211–212
  corrections applications, 213
  facial-scans, 211, 212–213
  finger-scans, 213
  iris-scans, 211, 213
  mug shots, 212–213
  probation and home arrest, 213
  surveillance, 165, 212
  technologies used in, 211
  typical deployments, 211–213
  vertical markets, 211–214
  voice-scans, 211, 213
layered biometric solutions, 132
legacy systems and citizen identification, 163–164
LFA (Local Feature Analysis), 70
liability for e-commerce/telephony, 198–199
limiting system access, 266
linear prediction coefficients, 90
live-scan, 117
live-scan devices, 116-117
live-scan fingerprint technology, 116
logical access systems, 14–15
logical and physical separation between data, 267

Index

low-quality enrollment, 17
LSIS (Lone Star Image System), 120

M

manually verify or determine identity, 9
match, 20, 21
matching, 20
match templates, 20
mature biometric applications, 146
measurements, 47
Mexican national elections, 72
Mexico
  border crossing, 230
  voter ID and elections, 216
Microsoft, 280
middleware
  authentication, 57
  PC/network access, 177, 178
military programs, 220
minutiae, 51, 52, 55–56
Mirage Resort, 72
mobile telephones, 89, 90
mug shots, 152, 212–213

N

nailbed identification, 113
nasal coarticulation, 90
national IDs, 215–216
national social security cards, 119
Nationwide Building Society, 82, 222
NCITS (National Committee for Information Technology Standards), 283-284
negative identification systems, 12–13
neural network, 71
New Delhi, India driver's licenses, 217

New York City Department of Corrections, 93
New York State Office of Mental Health, 56, 173, 227
Nigeria, 119, 215
NIST (National Institute of Standards and Technology), 283
noncommercialized technologies, 113
nonmatch, 20, 21
Novell Netware and PC/network access, 172
Nuance voice-scan and speech recognition technology, 93

O

Oceanside, California employee authentication, 219
O'Hare Airport, 231
one-to-one verification and keystroke-scans, 137
online banking, 223–224
optical technology and finger-scan, 54
opt-in *vs.* mandatory enrollment, 248–249
optional or mandatory enrollment, 271–272
Overt/Covert category, 256–258
overt *vs.* covert collection, 246–248

P

Panama voter ID and elections, 216
parallel layered solutions, 132
passive capacitance, 54
passports, 231

passwords, 3, 129
  ability to share, 4
  finger-scan systems, 48
  good, 4
  keystroke-scans, 139
  PC/network access, 180
  PINs, 4
  problems, 4
  simplicity of, 5
  tokens, 4
  universal, 5
pattern matching and finger-scans, 55–56
PC microphones, 89, 90
PC/network access, 145, 171
  API (application programming interface), 174
  authentication, 179
  Biometric Solution Matrix, 175
  costs, 177–178
  current applications, 172–173
  deployment issues, 178–180
  downturns in PC market, 174–175
  effectiveness, 175
  encryption, 174
  enrollment, 179
  exclusivity, 175
  facial-scans, 177
  fallback procedures, 179–180
  file formats, 174
  financial sector, 224
  finger-scans, 177, 178, 192
  future trends, 173–175
  healthcare, 226–227
  large-scale, 173
  legal issues, 174
  middleware, 177, 178
  multiple users, 179
  Novell Netware, 172

PC/network access *(continued)*
  password-authenticated users, 180
  PKI, 174
  privacy, 174
  receptiveness, 175
  related biometric technologies and vertical markets, 177
  resource access from remote locations, 179
  roaming users, 179
  scope, 176
  security, 180
  single sign-on applications, 175
  smart cards, 174
  stronger auditing and accountability, 174
  urgency, 176
  user orientation, 180
  voice-scans, 177
  Windows NT/2000, 172
performance
  highly controlled environments, 23
  real-world testing, 36
personal privacy, 243–244
personal storage *vs.* template database storage, 253–254
Peru voter ID and elections, 216
Philippines, 119
  benefits distribution, 218
physical access devices, 79
physical access systems, 14–15
physical access/time and attendance, 145, 180
  authentication, 186
  Biometric Solution Matrix, 184–185
  costs, 183
  current access system, 186
  current applications, 181–182
  deployment issues, 186–187
  effectiveness, 184
  exclusivity, 184
  fallback procedures, 186–187
  finger-scans, 183
  future trends, 182
  hand-scans, 183
  ingress/egress, 186
  iris-scans, 183
  receptiveness, 184
  related biometric technologies and vertical markets, 183
  retina-scans, 183
  scope, 185
  urgency, 184
  verification, 186
physiological biometrics, 10
physiological characteristics, 10
PINs, 3–4
  ATMs, 202
  problems, 4
PKI, 129, 174
platen, 47, 50
Plaza, 72
point-of-sale transactions, 5
positive identification systems, 12
postrelease programs, 93
potential, 264
presentation, 17
Printrak, 156
prisons, 82
privacy
  application-specific risks, 246–255
  biometric deployments, 238–239
  concerns, 239–246
  facial-scan abuse of, 75
  finger-scan, 60
  informational, 239–243
  invasive, 238
  keystroke-scans, 137
  PC/network access, 174
  personal, 243–244
  policies and protections, 272
  retina-scans, 111
  risks, 237–238
  surveillance, 165
privacy neutral, 238
privacy protective, 239
privacy sympathetic, 238
privacy-sympathetic data handling and citizen identification, 164
probation and home arrest, 213
protecting postmatch decisions, 266
public safety surveillance, 165
public sector *vs.* private sector, 250–251
Purdue Credit Union, 222

Q

quality enrollment, 17
quality score, 21

R

raw biometric data, 17
real-world FNMR (false nonmatch rates), 31–32
real-world performance testing, 36
receptiveness
  citizen identification, 160
  criminal identification, 155
  e-commerce/telephony, 195
  PC/network access, 175
  physical access/time and attendance, 184

Index

retail/ATM/point of sale (POS), 204
  surveillance, 166
remote student authentication, 232
remote transactions and increased trust, 191
response times for citizen identification, 163
retail/ATM/point of sale (POS), 144, 201
  1:N identification, 206
  authentication, 207
  Biometric Solution Matrix, 204–205
  check and credit card fraud, 203
  complexity of deployment, 203
  costs, 206–207
  current applications, 201–202
  customer motivation, 208
  deployment issues, 207–208
  effectiveness, 204
  exclusivity, 204
  existing system integration, 207–208
  facial-scans, 206
  fear of system errors, 204
  finger-scans, 202, 206
  future trends, 203–204
  iris-scans, 206
  process flow and ergonomics, 207
  receptiveness, 204
  related biometric technologies and vertical markets, 206
  scope, 205
  technology obsolescence, 208
  urgency, 205
retina-scans, 10, 17, 107
  accountability, 106–107
  components, 107

deployments, 110
difficulty of use, 111
distinctive features, 108–109
false matching resistance, 110
financial sector, 221
FNMR (false nonmatch rates), 32
FTE (failure-to-enroll), 35
identification, 13
image acquisition, 108
limited applications, 112
one-to-many identification, 109
physical access/time and attendance, 183
physiological trait stability, 110
privacy, 111
security, 106–107
strengths, 107, 110
susceptible to changes, 28
technology risk ratings, 258
template generation, 109
template matching, 109
user discomfort, 111
weaknesses, 107, 111
risk, 264
Royal Canadian Mounted Police, 212
RPM (Rapid Pay Machine) kiosks, 223

S

San Francisco Airport, 231
scalability and citizen identification, 162–163
scanner, 47
scope
  BioPrivacy Best Practices, 260
  citizen identification, 160
  criminal identification, 155

e-commerce/telephony, 195
PC/network access, 176
physical access/time and attendance, 184
problems, 149
retail/ATM/point of sale (POS), 205
surveillance, 166
score, 20–21
security, 4–5, 40
  e-commerce/telephony, 200
  education, 233
  financial services information, 281–282
  hand-scan systems, 105
  increased, 4–5
  keystroke-scans, 136–137
  PC/network access, 180
  retina-scans, 106–107
  universal passwords, 5
security cameras and surveillance, 165
security tools, 265–266
serial layered solutions, 132
shopping district, 75
short-time spectrum of speech, 90
signatures, 126
signature-scans, 10, 17, 123
  acquisition hardware, 124
  authorize transaction message, 124
  components, 124
  data acquisition, 125
  data processing, 126
  deployments, 127–128
  distinctive features, 126–127
  dynamics measured by, 126
  electronic tablet, 125
  enrollment, 130
  existing processes leverage, 128–129

signature-scans *(continued)*
  false matching, 127
  false nonmatches, 127
  financial sector, 221
  FNMR (false nonmatch rates), 30
  fraud reduction, 128
  FTE (failure-to-enroll), 35
  inconsistent signatures and increased error rates, 130–131
  limited applications, 131
  perceived as noninvasive, 129
  PKI, 129
  resistant to imposters, 128
  specific details recorded by, 126–127
  static signature, 125
  strengths, 123, 128–130
  technology risk ratings, 258
  template generation, 127
  template matching, 127
  users ability to change signatures, 129–130
  users unaccustomed to signing on tablets, 131
  verification, 130
  weaknesses, 124, 130–131
silicon technology and finger-scans, 54–55
single false match rate, 26–27
single FNMR (false nonmatch rates), 32
single FTE (failure-to-enroll), 37–38
slap, 116
smart cards, 174, 203, 244
South Africa benefits distribution, 218
Spain benefits distribution, 218
SpeakerKey technology, 93

spectograms, 90
speech recognition, 87, 94
speech recognition systems, 88
static signature, 125
Stratosphere Hotel and Casino, 72
students, 251–253
student services, 232
stylus-operated PDAs, 125
subcutaneous hand-scan, 113
Super Bowl, 273–275
Supermarket Banks, 224
surveillance, 72, 145–146
  automated matching with cameras, 165
  Biometric Solution Matrix, 166
  casino usage, 164
  costs, 167–168
  current applications, 164–165
  deployment issues, 168–169
  deterrence *versus* detection, 169
  effectiveness, 166
  exclusivity, 166
  existing acquisition hardware, 168
  facial-scans, 167
  future trends, 165
  intervention process, 169
  law enforcement, 165, 212
  performance, 165
  privacy, 165
  public safety, 165
  receptiveness, 166
  related biometric technologies and vertical markets, 167
  scope, 166
  security cameras, 165
  target database size and quality, 168

  urgency, 166
  voice-scans, 167
system false match rate, 26–27
system FNMR (false nonmatch rates), 32
system FTE (failure-to-enroll), 37–38
system potential capabilities evaluation, 263–264
system termination provisions, 265

T

Taj Mahal, 72
Takefuji Bank, 82, 222
telephone transactions, 224
telephony-based voice-scan implementations, 94
template creation, 16–18
template matching
  facial-scans, 64, 68–69
  finger-scans, 52–53
  hand-scans, 102
  iris-scans, 82
  keystroke-scans, 135–136
  retina-scans, 109
  signature-scans, 127
  voice-scans, 92
templates, 18–20, 245
  comparison, 22
  distinctive features, 19
  enrollment, 20, 21
  facial-scans, 68
  file size, 18
  fingerprint, 114
  finger-scans, 52, 120
  hand-scans, 102
  incorrectly judged, 27–28
  iris-scans, 81
  keystroke-scans, 135–136, 137
  match, 20, 21
  matching, 20

Index

proprietary, 19
retina-scans, 109
signature-scans, 127
unique, 19
verification, 20, 21
voice-scans, 91–92
template storage *vs.* identifiable data storage, 255
thermal facial-scan, 113
thermal imaging and finger-scan, 55
third-party auditing and oversight, 269
threshold, 20, 21
Timemac Solutions, 93
T-NETIX, 93
tokens, 4, 129
touch typists, 135
trabecular meshwork, 80
traditional authentication technologies *vs.* biometrics, 3–5
transactional revenue models, 191
travel and immigration, 210, 228
  AFIS, 229
  air travel, 229–230
  border crossing, 230–231
  employee access, 231
  finger-scans, 229
  hand-scans, 229
  iris-scans, 229
  passports, 231
  technologies, 229
  typical deployments, 229–231
travelers, 251–253
true verification attempts, 36
Trump Marina, 72
trusted parties and e-commerce/telephony, 192
typing patterns, 135

U

Uganda, 75
  voter ID and elections, 217
ultrasound technology and finger-scan, 55
unenrollment, 267–268
unique identifier, 262–263
United States
  benefits distribution, 218–219
  driver's licenses, 217
universal passwords, 5
University of Georgia, 232
unsafe neighborhoods, 72
urgency
  citizen identification, 160
  criminal identification, 155
  e-commerce/telephony, 195
  PC/network access, 176
  physical access/time and attendance, 184
  retail/ATM/point of sale (POS), 205
  surveillance, 166
user ownership *vs.* institutional ownership, 253
users
  control of personal data, 267–268
  feedback, 29
  presentation changes, 29

V

vein identification, 113
verbal account authentication, 94
verification, 5, 12–13
  anonymous, 268
  appropriateness of, 13–14
  BioPrivacy Impact Framework, 249–250
  disclosures, 271
  hand-scans, 101
  iris-scans, 82
  keystroke-scans, 134–135, 137
  physical access/time and attendance, 186
  process disclosure, 272
  signature-scans, 130
Verification/Identification category, 256–258
verification systems, 10, 12
  accuracy, 13
  false nonmatch, 27
  FMR (false match rate), 24
  match/no-match decision, 13
  PC and network security, 13
  speed, 13
verification templates, 20, 21
  feature analysis, 71
vertical markets, 209
  education, 232–233
  financial sector, 210, 220–225
  gaming, 232
  government sector, 210, 214–220
  healthcare, 210, 225–228
  law enforcement, 210, 211–214
  travel and immigration, 210, 228–231
Virgin Airways, 83, 230
Visionics, 70
Visionics facial-scan technology, 72
voice-scans, 10, 17, 87, 88
  acquisition devices and ambient noise on accuracy effect, 95
  components, 88
  data acquisition, 89–90
  data processing, 90
  deployments, 92–93
  distinctive features, 90–91

voice-scans *(continued)*
  financial sector, 221
  FNMR (false nonmatch rates), 30
  FTE (failure-to-enroll), 35
  lack of PC usage suitability, 96
  large template size, 96–97
  law enforcement, 211, 213
  leveraging existing telephony infrastructure, 94
  low accuracy perception, 96
  negative perceptions lack, 95
  PC/network access, 177
  performance, 89
  proprietary software engines, 88
  resistance to imposters, 94–95
  speech recognition and verbal account authentication synergy, 94
  strengths, 87, 93–95
  surveillance, 167
  technology risk ratings, 258
  template creation, 91–92
  template matching, 92
  temporal changes, 28
  user feedback, 29
  weaknesses, 88, 95–97
  voluntary system usage, 267–268
voter fraud, 75
voter ID and elections, 216–217
voter registration
  citizen identification, 159
  voting, 157
Voxtron, 93
VPN (virtual private network), 174, 179

W

Web sites
  biometric authentication, 190
  biometric functionality, 172
Welsh Valley Middle School, 57–58, 232
Westernbank, 56, 222
Windows NT/2000 and PC/network access, 172

X

X9, 281
X9.84, 174, 220, 281–282

Y

Yemen national IDs, 215